ENCHANTMENT AND EXPLOITATION

The Life and Hard Times of a New Mexico Mountain Range

William deBuys

UNIVERSITY OF NEW MEXICO PRESS
Albuquerque

Design by Milenda Nan Ok Lee

Library of Congress Cataloging in Publication Data

DeBuys, William Eno.
Enchantment and exploitation—the life and hard times of a
New Mexico mountain range.

Bibliography: p.
Includes index.
1. Sangre de Cristo Mountains (Colo. and N.M.)—History.
2. Man—Influence on nature—Sangre de Cristo Mountains
(Colo. and N.M.) 3. Environmental policy—Sangre de Cristo
Mountains (Colo. and N.M.) I. Title.
F802.S35D43 1985   978.8'49   85-5833
ISBN 0-8263-0819-8
ISBN 0-8263-0820-1 (pbk.)

First edition

to
my mother
who especially appreciates a good book
and to my father
who especially appreciates a day in the good outdoors

Contents

CONTENTS

Maps

Illustrations

CONTENTS

LIST OF ILLUSTRATIONS

CONTENTS

Tables

Preface

THERE IS A BRIEF PERIOD in late fall when the southern Rocky Mountains reveal themselves more fully than at other times of year. It comes after the dazzling gold of the aspen and all the other autumn colors have faded to dull tones, and it precedes the deep winter snows that cover the high country with a white mask. At such a time one November I hiked into the Pecos Wilderness in the Sangre de Cristo Mountains of Northern New Mexico. On this trip, as on others, I had gone into the high country both to enjoy the land and to learn from it.

At that time of year the mountain forests of spruce and fir wear their blackest shade of green, and the ceaseless rush of the wind suggests that winter's first blizzard may not be far away. On foot and alone, I set out for the high country, and after a long day of climbing I spent a twenty-degree night beside the East Fork of the Santa Barbara, a clear and ice-hung stream that drains some of the most remote and least visited land in the range. In the morning, climbing again, I found occasional signs of animal life in the early-season snow that lay in the forest. Here a coyote had trotted beside the river. There an elk had browsed a shrub. I was surprised to find the track of a snowshoe hare that had dashed through the trees, very erratically, and then simply stopped. I looked closer and found a splash of red in the snow, and beside it, the trident-shaped clawprint of a goshawk or a great horned owl.

I walked on and, not much farther, suddenly caught sight of a pyramid of crumpled metal a short distance off the trail. It was the wreckage of an airplane. I was not immediately alarmed because downed aircraft are not uncommon in the mountains. But I had never heard of a wreck on the East Fork. I began to feel an adrenal rush as I approached the plane, noticing first a torn seat cushion with its stuffing still unmolested by wood rats. Behind the wreckage I saw a broken tree with fresh sap flowing from the wound, and closer, a litter of tissues, a cowboy hat, a glove.

I found four bodies in the plane, all quite frozen, and I was relieved to see no sign of scavengers or even of struggle or suffering. Apparently death had been instantaneous. I had no idea how long ago it had occurred, and I stayed near the wreckage only long enough to copy down the identification number of the plane and to be sure that its occupants were indeed dead and that there was nothing I could do.

Clearly there was nothing, but it seemed that the groaning of the wind grew steadily louder in the time I remained near the wreck. I struggled to think clearly whether there was not some duty I should perform, and I began to doubt whether I could be of any service to the dead, even as a messenger carrying word of their misfortune back to civilization. Frozen in their aluminum tomb and shrouded by drifts of snow, they seemed well buried. I remember wishing I had an offering of cornmeal or pollen or holy water with which to sanctify the place, not because of what had befallen them, but because my anxious presence had profaned their place of rest and ultimate quiet. The mountains often inspire feelings of personal insignificance but I had never felt them so intensely before. Here were mortality and finality, and I was a nervous intruder.

The official investigation of the crash eventually revealed that the four individuals who lost their lives in the canyon of the East Fork had come to the mountains with too little understanding of the mountain environment and too little respect for the perils their ignorance posed for them. They had come as sightseers, fascinated by the awesomely glaciated landscape, but unaware that the rubble-sided peaks rose more steeply than their frail plane had the power to climb. Perhaps they looked down on a band of elk in one of the grassy parks. Perhaps they flew wing to wing with a golden eagle or a migrating rough-legged hawk. Too late, the pilot realized how sharply the headwall of the canyon rose to the sky. At the last instant he banked his plane into a tight turn. The lower wing struck the top of a spruce and pitched the craft into the ground. A day later came the first snow of the year, and the white plane, its emergency transmitter broken, was absorbed into a white and frozen world.

The point of this story is deceptively simple. It has nothing to do with fake profundities about technology or wilderness, and still less with advice on air travel. It is simply this: in an unforgiving environment, small errors yield large consequences.

This minor yet muscular truth characterizes every pioneer experience, and it is one of history's themes in New Mexico. Centuries of human experience afford abundant examples of small miscalculations leading to large-scale misfortune. One hundred years ago, for instance, a great many relatively small decisions by stockmen to increase the size of their herds resulted in a debacle of overgrazing that seriously undermined the pastoral economy of the mountain villages.

The trick of living in the mountains begins with understanding the power of the landscape and the limits it imposes. By extension, the region's history begins with the story of how people have learned that lesson—and at times forgotten it.

Acknowledgments

WHEREAS MANY WRITERS acknowledge the assistance of the National Endowment for the Humanities or some other beneficent institution, I have a somewhat different debt of gratitude to express. Although I did not know it when I found the crumpled wreck and its lifeless passengers on the East Fork of the Santa Barbara, a reward had been offered for the discovery of the lost airplane, and eventually I received a large portion of it. My share of the reward, which I was acutely aware of having in no way earned, enabled me to continue work on this book, then in its earliest stage. While the reward gave me the means to keep writing, it also, I felt, gave me an obligation to persevere in a project that I might otherwise have abandoned. I felt that in a sense I had received a "grant" from the mountains themselves and that I was obliged to give something back. This book is what I give.

Somewhat more concretely, I am also deeply indebted to the Graduate School of The University of Texas at Austin, which supported me while I wrote the final drafts of the manuscript, and the John Muir Institute for Environmental Studies of Napa, California, which funded several months of research on range management in northern New Mexico.

Among the many individuals on whose assistance I repeatedly relied, I should like especially to thank Henry Carey and John Donald, who generously tutored me on various aspects of resource management and whose friendship was one of the primary rewards of this undertaking. For similar reasons I am also indebted to Pete Tatschl of the U.S. Forest Service and to many of his colleagues at District, Forest, and Regional levels; to Bob Lange, formerly of the New Mexico Game and Fish Department; and to Malcolm Ebright of the Center for Land Grant Studies. And I thank John Cotter for the wonderful maps.

The people who contributed most to this book, however, are the people of El Valle and neighboring villages—the Romeros, Montoyas, Aguilars, and many others. I wish especially to thank Jacobo and Eloisa Romero and Tomas Montoya, who embody everything that is meant by the phrase *buen vecino*. Without them, there would be no book, and I earnestly hope that this volume meets with their approval.

Every writer needs imaginative and sympathetic critics, and I was fortunate to have two who were especially so: Peter Decker, Colorado's most literate and literary rancher, and William H. Goetzmann of The University of Texas at Austin, who has few if any peers as a scholar, writer, and teacher of Western history. Another peerless friend to whom I am greatly indebted is Beth Hadas of the University of New Mexico Press, who is largely responsible for bringing this book to press.

Finally, I would like to thank my parents and sisters for their patience and support; Alex Harris for his confidence that something would come of this and his many acts small and large that helped ensure that something did; and most of all, my wife, Anne, for her encouragement, her insight, and her faith. No one has had better company on a hard trail.

ENCHANTMENT AND EXPLOITATION

Introduction: Place

To the east of the Villa, about one league distant,
there is a chain of very high mountains which
extends from south to north so far that its limits
are unknown.
                                    —*José de Urrutia, 1766*

THE ROCKY MOUNTAINS, spine of the continent, give birth to
the Rio Grande in southern Colorado and fork to either side.
In the west they become the San Juan Range, which dimin-
ishes from its Colorado heights to a long tangle of ridges in
New Mexico. East of the river the mountains form a rugged
sierra walling the grassy sea of the Great Plains from the increasingly arid country of the intermountain West. The exotic
name of these mountains, Sangre de Cristo—Blood of Christ,
derives, according to one legend, from a distant time when
Pueblo Indians slaughtered the Spanish missionaries who had
come to save them from the perdition of their native paganism.
One of these missionaries, at the moment of his martyrdom,
prayed to God for a sign that his death would not be in vain,
and as though in response, the snowcapped peaks of the sierra
were suddenly bathed in a vivid crimson—they glowed with
the Blood of Christ! The legend gives no hint as to how word
of this miracle got out (and the question is an interesting one
since the one believing witness to the event was instantly
hacked to bits), but the commonplace of sunsets reddening
the Sangres helps to keep the tale alive.

[3]

There exists a second and more credible history of the name. Long known to Spanish colonists simply as the Sierra Madre, the mountains are broken by a relatively accessible pass near present-day Walsenburg, Colorado. As early as 1779 the river flowing east from the pass was identified on maps by the name Sangre de Cristo. The name was subsequently applied to the pass itself, and later, as sizable numbers of traders and trappers entered New Mexico by way of the pass, to the entire mountain range. This last extension of the name was firmly cemented in the public mind by the florescence of the passionately religious Penitente Brotherhood during the middle and late 1800s. This sect, whose rites featured bloody self-torture and flagellation, became famous for its yearly re-enactments of the original shedding of the Sangre de Cristo. It is poetic justice that the mountains where the Penetentes were and are still strongest came to bear the name of their extravagant obsession.

The Sangres stretch farther south than any other spur of the Rockies and yield to a landscape of mesas and plains not far from Santa Fe, at the small, tough town of Pecos, New Mexico. You could hardly say that the Rockies whittle down to a mild end. The mountains above Pecos and Santa Fe soar to altitudes of over 13,000 feet, and their canyons and ridges are home to elk, deer, bighorn sheep, black bear, and cougar. Almost all of the mountain country is now included in the Carson and Santa Fe National Forests, and although numerous small farming and ranching communities lie scattered through the lower elevations, higher up, where the two national forests jointly manage the Pecos Wilderness, the land is as free from sustained human impact as any land in the Southwest. There in the roadless country where, according to local wisdom, you get ten months of winter and two of very late fall, a dozen of New Mexico's tallest peaks rise treeless out of the forest. In the giant bowls, or cirques, that glaciers carved from the bases of the peaks, spring the headwaters of the nine-hundred-mile Rio Pecos as well as tributary sources of the nearby Rio Grande and the distant Mississippi.

Like any mountain range, particularly one that divides the waters of major river systems, these mountains are collec-

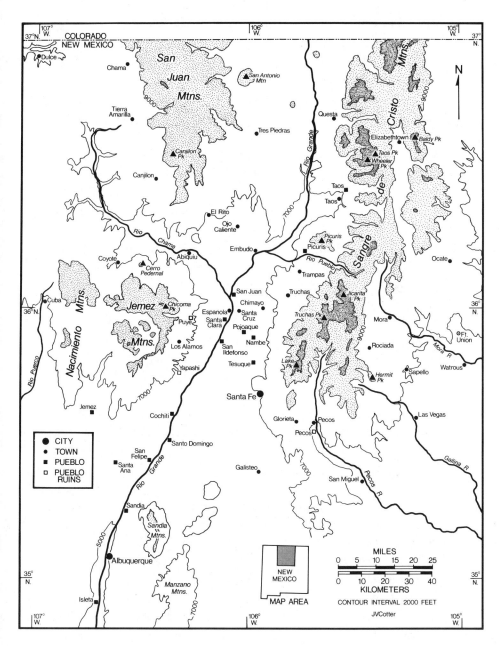

Map 1. Landforms and settlements of north central New Mexico.

tively a kind of pivot in the natural history of the region. The land below them is made from their eroded wastes; they change and channel the weather, dividing one environmental region, one flora and fauna, from another. Equally, as a barrier to communication and conquest and as the source of valuable water, timber, game, and grazing, they have played a pivotal role in human affairs.

And what remarkable affairs they are. The history of New Mexico is as rich as it is long. Anyone who studies it is at pains to remember that New Mexico has been part of the United States for only a century and a quarter and that for nearly twice as long it belonged to Spain and Mexico. And before the coming of the Spanish, for much longer, it indisputably had belonged to native, village-dwelling agriculturalists.

The mountains offer a good vantage from which to view the history of the region. They fill the triangle formed by the towns of Santa Fe, Taos, and Las Vegas, New Mexico. Within this triangle, most of the decisive events in New Mexico's early history took place. The mountains were a valued goal to some of the people who came to them, and an obstacle to others. In every case, they were always an actor as well as a stage, for their influence reached to the heart of every enterprise.

The Sangre de Cristo Mountains recommend themselves for study because no other region in all of North America so richly combines both ecological and cultural diversity. The range stands with its feet in the near-desert of the arid Rio Grande Valley and its head in the alpine tundra of thirteen-thousand-foot peaks. Between those extremes lie as many as nine or ten discrete ecological zones, depending on how one classifies them. Even more impressive, however, is the mix of cultures one finds in the Sangres. The Pueblo Indians and their ancestors have made their homes here for at least two thousand years, and for the last four hundred they have shared their land with the sons and daughters of Spanish-speaking pioneers from Mexico. The Hispanic *pobladores* of northern New Mexico built their simple adobe villages in scores of irrigable valleys nestled in the mountains' lower slopes, and

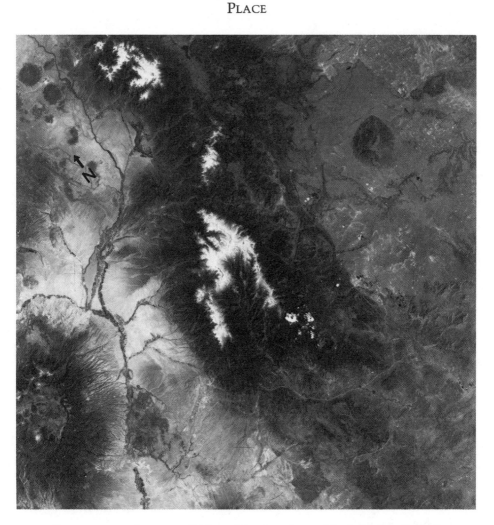

*Photo 1. North central New Mexico from Skylab, June 1973.
Note snow above timberline in the Sangre de Cristo Range
and irrigated land along Rio Grande and other rivers. The
rounded, dark areas are volcanic formations, largest of
which is the Valle Caldera of the Jemez Range in lower left.
Courtesy of Earth Resources Data Center, Sioux Falls, S.D.*

there these remarkable communities remain today, isolated, weatherbeaten, and, to most Anglo-Americans, quite foreign.

My first introduction to the Sangres was as a research assistant to a social scientist writing a book on certain aspects of the region's cultures. I was given license to pursue my research however I desired—whether in saloons or elementary schools did not matter, so long as I conveyed such findings as I made to my employer. Unfortunately, this altogether ideal arrangement ended in failure. I found it impossible to write intelligently about the people of the region when I knew so little about the demanding physical environment in which they dwelled. I realized that until I understood something of the influence of the land, I could not begin to understand the people whose lives and history bound them to it.

This book grew from an opportunity, years later, to build upon that intuition. It is not a "regional study" in the traditional geographic sense. The concept of region is virtually impossible to define with precision, and I have no wish to argue that the Sangres constitute or even typify any particular geocultural area. Rather, what I have sought here is to capture a sense of the mountains' being, a sense, that is, of place.

In studying place I have wanted to show that a society's relationship to the environment is reciprocal: it both changes the physical world and is changed by it. I have also attempted to show that the study of those changes can illuminate, in a most appropriate way, the main themes of the cultural history of New Mexico.

One theme to which I have given particular attention concerns the competition among the region's diverse cultures for limited resources of water, game, wood, and grazing, and this competition is virtually as intense today as it was two thousand years ago, particularly if one adds "employment" to the list of resources at stake. My hope is that by clarifying the issues of the past, this book will contribute to informed debate on the present use and management of resources in northern New Mexico. In both a literal and a figurative sense, its purpose is to reveal the common ground.

I should point out that this book presents no argument for

geographical determinism, as did, for instance, Ross Calvin in his 1934 study of the Southwest, *Sky Determines*. Rather I mean to emphasize that in adapting to the environment, societies change it both purposefully and by accident, and in turn adapt to the changes they have wrought—sometimes by changing the environment still further, which is the Anglo way, the way of dams for irrigation and more dams for desalinization. At every step in the process of adaptation and change there are opportunities for choosing between alternatives, and the choice by a people of one alternative over another depends in large measure upon their outlook and priorities—in a word, upon their culture.

Culture provides a filter through which people perceive the environment around them and their relation to it. It screens out certain influences or possibilities while allowing others to pass on to awareness, sometimes greatly intensified. For the Pueblos the landscape was manifestly spiritual in certain locations and at certain times. The mountain peaks were sources of spiritual energy; the lakes were doors to the underworld. The land and sky were living things which the Pueblo people supplicated through elaborate ritual to ensure the orderly progression of the seasons and the stability of their communities.

To the Hispanos of New Mexico the land was not a sacred thing as it was for the Pueblos but it was still not altogether inanimate. It was the mother and protector of their traditional subsistence pastoralism. In many cases it was a communal thing, belonging not to individuals but to whole villages as a collective possession.

Initially in the Anglo view the land possessed neither spiritual nor communal qualities. Land was simply a commodity, which like wheat or iron ore might be advantageously bought and still more advantageously sold. At the end of the nineteenth century this view yielded somewhat to a kind of scientifically oriented communalism that argued that certain lands, particularly the high mountain forests, should be held in trust for all members of society in order to protect water quality, timber supplies, and other resources. The Forest Reserves, later renamed National Forests, were the result. Never entirely separate from this communalism was the notion that the wild country contained in the National Forests possessed a spiritual quality of vital importance to our national way of life. Eventually that idea was elaborated in a system of National Wilderness Areas, of which the Pecos Wilderness in the southern Sangres is one of the best known.

The story that follows has been divided into two "Books," the first of which examines northern New Mexico's continually changing character as a frontier for its three principal cultures. Readers who are already well versed in the early history of New Mexico may wish to move on rather quickly to the more intricate story contained in Book Two, but I urge them to pause to assess what they will miss. This book would not adequately fulfill its purpose of contributing to an informed debate on contemporary issues if it were written solely for a small professional audience. Rather, its story is addressed to everyone with an interest in the region's unique sense of place, which the centuries of change and conflict summarized in Book One have largely shaped.

The first actors to leave their imprint on the mountain stage were the hunters and gatherers of the Oshara tradition. For reasons known only to them, they made summer hunting camps on the highest and windiest of the alpine divides. Their descendants, the Pueblo Indians, also lived in the shadow of the mountains and found in the physical landscape a spiritual reality more elaborate than that of any other Southwestern Indian group. They were joined in the mountain country by the Jicarilla Apaches of the eastern foothills, and occasionally, and rarely peacefully, by the Comanches of the eastern plains.

The first of two great revolutions in New Mexican affairs came with the Spanish Entrada of 1548 and Spain's subsequent colonization of New Mexico. Suddenly the prehistoric world of the Southwest became historical. Initially the Spanish-speaking *pobladores*, the colonists and populators of an empire, were merely the acolytes of a frantic search for mineral treasure, but as the colony's dreams of wealth and glory faded, they became the principals in one of Spain's most audacious and least successful efforts to spread Iberian civilization through the New World. Up the long trail from Mexico they came, carrying in their veins not just the blood of Old World Spain but the blood of Indian Mexico as well. They settled into the valleys and edges of the mountains and fashioned a new cultural identity from their struggles there; they became *New Mexicans*. These Hispanic Southwesterners (who from their point of view are *norteños*, or northerners) are seldom mentioned in the books that present American history to schoolchildren, but their story is as much a part of the North American colonial experience as that of the Pilgrims or the Virginia planters.

After the Comanche wars, the frontiers of Hispanic New Mexico began slowly to expand, ushering in a new period of discovery and cultural growth. Almost immediately, however, the Hispanic frontier collided with the still more vigorous frontier of Anglo-American expansion. This was the second of New Mexico's great revolutions. The trappers and merchants who trekked across the Great Plains from St. Louis to New Mexico were men still looking for their country and still finding and stealing it. They were followed by ranchers and farmers, miners, lawyers, and thieves, soldiers, and tourists who came piling into New Mexico in wave after wave, and at least part of every wave broke against the mountains, leaving an indelible imprint.

Book Two begins with a subtle shift in the tide of Anglo conquest. The presence in the mountains of parties of the Wheeler Survey, inventorying the natural resources of the region, presaged the rise of scientific and entrepreneurial approaches to land use, strongly aided and subsidized by the federal government. This period also saw a radical disjuncture in the natural history of the region as the exploitation of the land reached a new and unsustainable intensity. In a relatively short span of time, elk and bighorn sheep were exterminated in their mountain habitat, the last refuge of a formerly much broader territory. The local extinction of grizzly bear, ptarmigan, and pine marten shortly followed. Grasslands were grazed to exhaustion, then forests burned to create more grasslands, which in short order were also overgrazed. Plant ecologies were ruptured by overuse and then re-formed, often with newly introduced species replacing beleaguered native ones. In many areas erosion sapped the strength of the soil, and arroyos lowered the water table, preparing the way for the alternate miseries of drought and flood. As the land grew weaker and poorer, the vitality of the human economy in and around the mountains also began to decline. In response, New Mexicans attempted, at first slowly and falteringly, to seek new balances in their use of the land, balances which are still evolving.

The aim of this book is to understand those balances, both past and present, and to explore the paradoxical human affairs that produced them. This is an auspicious time to undertake such an exploration, for as a result of environmental legislation passed in the 1970s, federal agencies are today obliged to permit a much higher level of public participation than ever before in the management of the lands they control. Considering the vast areas of northern New Mexico held by federal agencies, particularly the U.S. Forest Service, this requirement represents a significant opportunity for the people of the region. It means that in their collective search for wise use of their region's resources, they have the means and the right to make their views heard, if only they will use them.

These opportunities, discussed in the final chapter of this book, have the potential to broaden the meaningful expression of democracy in New Mexico and perhaps in a small way to heal some of the injury that has been inherited from the past.

The requirement that the public have an opportunity to participate in land-use planning acknowledges a lesson that the mountains of northern New Mexico have taught since time immemorial. Whether as sacred space, communal inheritance, or private property, the mountains have demonstrated again and again that the life of societies and the life of the land cannot be separated. What follows is an account of the linkage between those two lives, between human history and natural history.

BOOK ONE

THREE FRONTIERS

1

Spirit Homes

*To say "beyond the mountains" and to mean it, to
mean, simply, beyond everything for which the
mountain stands, of which it signifies the being.*
                                        —N. Scott Momaday

SMALL RED LIGHTS BLINK through the darkness all night above
Albuquerque and El Paso, above Denver and above every city
large and rich enough to buy a crest of high land, build trans-
mitting towers, and spray itself with an ether of television
and radio signals. We moderns do not have sacred mountains
in our lives, but we do have needs. Cut off the power to the
electronic installation atop the Sandias; turn off the transmitters in any city or town in the country and rapidly the
pattern of daily life will erode; people will feel disoriented, as
though an important part of the landscape had vanished.

The Pueblo Indians would understand. For centuries they
have known—and in spite of the decline of subsistence farming and the modern intrusions of tourism, Wonder Bread, and
the Bureau of Indian Affairs, it appears that they still do know—
what it is to depend on a mountain. In the cradle of the upper
Rio Grande, where the peaks of the southern Sangres tower
along the eastern horizon, the Pueblo homeland is shaped by
the mountains that surround it. For untold centuries, the horizon circle of blue mountains has been the image in every
individual's life of the limits of knowing and being and imag-

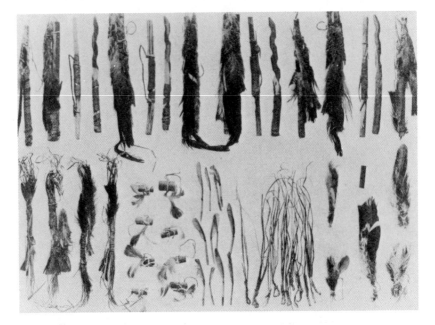

Photo 2. Prayer sticks collected from a shrine on the summit of Chicoma Peak by William B. Douglass, a surveyor for the General Land Office, probably in 1911. The sticks consisted of short willow or cottonwood twigs painted with the colors of the gods to whom they were addressed. Their message was encoded in the feathers, carvings, and sprays of grasses and flowers with which they were adorned. From Records of the Past, *1912, courtesy Duke University Library.*

ining. The mountains, as the rim of a sacred world, have been the collectors and transmitters, not of radio waves, but of the blessings and prayers of the cosmos.

In the Pueblo homeland, to borrow a phrase from Willa Cather, the sky is not so much the roof of the earth as the earth is a floor for the sky. Aridity, more than any other single factor, shapes this stark world, and the Pueblos through their centuries of experience have come to view rain as the essence of life and the thunderclouds hanging on the mountain peaks as gifts from the gods. But aridity, like any constant hardship, can be taken into account. What makes the Southwestern environment particularly difficult for subsistence farmers is less the lack of rain than the capricious timing of it. The summer thundershowers, which make the difference between crop failure and abundance, are notoriously violent and local. One farmer may have his fields drenched so rapidly that more water runs off than soaks in, while his neighbor's fields, less than a mile away, bake in the dizzying sun. Add to this the unpredictability of drying winds and of a growing season whose length may vary as much as two months from year to year, and one understands why the Pueblos have evolved a religion whose purpose is to control the environment.

In spite of the unpredictability of their environment, the Pueblos long ago came to believe that the cosmos was, above all, orderly. They viewed it as a place where everything animate and inanimate, physical and nonphysical depended upon every other thing in a kind of all-encompassing spiritual ecology. They believed that the cosmos, being orderly, was knowable, and being knowable, was also controllable. Their religion, based on these precepts, became a striving for the control over their environment that would ensure them an abundance of crops and game, victory over their enemies, and safety from witchcraft and disease.

The Pueblos held these beliefs in common, but they were not one people. They were united only by the broad outlines of their agricultural way of life, not by loyalties or even language. The westernmost among them inhabited the pueblos of Hopi and Zuni, each large and independent and each possessing its own language. Farther east, the Keres-speaking Pueblos lived at Acoma and Laguna in the dry mesa country west of Albuquerque, and also at Zia, Santa Ana, San Felipe, Santo Domingo, and Cochiti, clustered near the Rio Grande. A still larger and more heterogeneous group of pueblos lay scattered north and south along nearly two hundred miles of the great river the Spaniards first called Río Bravo del Norte.

These, the Tanoan Pueblos, are related linguistically to each other, and can be further subdivided into Tiwa, Tewa, Towa, and Tano. The Tiwa include Taos and Picuris in the north and Sandia and Isleta in the south, and their separation into four individual pueblos is so ancient that their languages are not mutually intelligible. The Tewa, on the other hand, appear to have dispersed more recently since speakers of one of its dialects can, with practice, effectively communicate with speakers of another. Today the Tewa comprise six pueblos: San Juan, Santa Clara, San Ildefonso, Nambe, Tesuque, and Pojoaque. The Towa, meanwhile, have only one village, which is Jemez Pueblo in the mountains west of the Rio Grande, but they formerly inhabited the once powerful and populous pueblo of Pecos, at the southern end of the Sangres, until its abandonment in 1838. By then, the Tano had likewise abandoned their pueblos and joined other groups.

The Pueblos of today remain a considerable mystery to the Anglo nation that surrounds them. Having learned reticence from centuries of conflict with the Spanish, they have also learned to be tight-lipped with curious tourists and social scientists, all equally eager to *know* the Pueblos but few of them able or willing to repay their informants with anything as valuable as the knowledge they wish to take away. There have been, of course, exceptions. One of the earliest Southwestern ethnologists, Frank Hamilton Cushing, for instance, became a close friend of the Zuni and staunchly defended them through difficult legal and political battles, but even he, establishing an all too often repeated pattern, ultimately alienated his Indian friends by publishing information given to him in confidence. While some anthropologists have been more faithful than Cushing, many others have been less so.

It should come as no surprise, therefore, that the Pueblos, who have had as much experience with anthropologists and ethnologists as any people on earth, continue to be reluctant to share their private beliefs with the many strangers, academic and otherwise, who come seeking them. Even within pueblos there is strict control over the possession of sacred

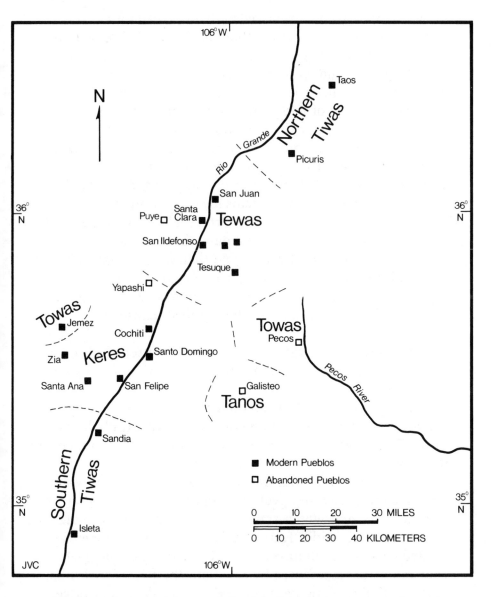

N

Taos

Northern Tiwas

Rio Grande

Picuris

San Juan

36° N

Santa Puye Clara

Tewas

San Ildefonso

Tesuque

Yapashi

Towas Jemez

Cochiti

Towas

Pecos

Zia Keres Santo Domingo

Pecos River

Santa Ana San Felipe

Galisteo

Tanos

Sandia

■ Modern Pueblos
□ Abandoned Pueblos

Southern Tiwas

35° N

Isleta

0   10   20   30 MILES

0   10   20   30   40 KILOMETERS

JVC   106°W

36° N

35° N

Map 2. Indian pueblos and Pueblo language groups of north central New Mexico.

knowledge so that unauthorized individuals cannot violate the rules of secrecy. As a result, there is a dearth of reliable information about many facets of Pueblo life, and even the most knowledgeable students of Pueblo culture, some of them Pueblos themselves, have had occasion to confess ignorance of important aspects of Pueblo world view and religion.

Accordingly, it is with trepidation that I sketch here the relation of the Rio Grande pueblos to the high mountains that physically and spiritually enfold them. What follows is at best an approximation, though a sympathetic and, I hope, an accurate one, of what Pueblo Indians have traditionally believed. It is drawn in part from Alfonso Ortiz's *The Tewa World* and so is less true for other Pueblo groups than for the Tewa, and less true for any other Tewa pueblo than for San Juan, where Ortiz did his work. Nonetheless, in a general sense it is faithful to an attitude shared by all Pueblos, an attitude of deep veneration and intense delight in the physical landscape.

Their world has a center that is like the center of a human being. It is a navel in the earth, taking the form of a shallow depression in a dance plaza at the center of the pueblo. Called the mother-earth navel, it appears to be only a dusty basin where rain might collect, but it is also a door that leads directly to the Below, where the Pueblo people first lived and where their most important deities still dwell.

Radiating outward from the pueblo are other shrines which are linked to the earth navel by a continuous flow of prayers and blessings and which transmit and receive the messages of the spiritual world. Four of them, marked by stones, are located close to the four ceremonial entrances to the pueblo. Four more are represented by the Tsin, the four sacred hills. Pueblos tell their children that these hills are dangerous places, riddled with tunnels, where boogiemen called the Tsave Yoh will take them if they misbehave. The Tsave Yoh, who are also healers of the sick and enforcers of pueblo law, live on the Tsin along with other important spirits. The hill of Tsi Mayoh, from which the Spanish village of Chimayo takes its name, is considered a Tsin by several pueblos, and so is Tun

Yoh, or Black Mesa, which rises abruptly beside the Rio Grande between the pueblos of Santa Clara and San Ildefonso.

Four is the most sacred number. Most dances are repeated four times so as to bring all of the Pueblo world, all four directions, into the sweep of their authority. In the dim time just before the Pueblo people emerged from the Below, pairs of brothers went out in each of the four directions to explore the earth and determine if it were ready for men to live on. The brothers who went north were blue, those to the west yellow; red ones went south and white ones east. They all came back with the same report: the earth was still too soft, too uncooked, for them to travel far. Where they stopped, they took up a handful of mud and flung it in the direction they had been traveling. Thus the Tsin were created. Each pair also reported that the unfinished earth was still too hazy a place for them to see far, but away in the distance they could make out the outline of a tall mountain. These were, and are, the sacred mountains of the four directions.

In the west, powerful from its nearness, stands Chicoma, the tallest mountain of the Jemez Range, which is sacred not just to the Tewa, but to Taos, Jemez, Cochiti, and others. The mountain of the south, farther away, is called Turtle Mountain by the Tewa because of the humpbacked shape it presents to the north. Most New Mexicans know it as Sandia Crest, which towers over Albuquerque.

About the mountain of the north there exists some confusion, at least among anthropologists. Ortiz argues that it is Canjilon Mountain, in the Chama watershed northwest of San Juan, while others, among them such pioneers of southwestern ethnography as John Peabody Harrington and Edgar Hewett, have reported that the sacred mountain is San Antonio Peak, an ancient shield volcano that rises in spectacular isolation next to the Rio Grande near the Colorado border.

The high peaks of the Sangre de Cristo Range bound the Pueblo world on the east, and because of the nearness of the mountains and the abundance of high craggy summits, different pueblos hold different mountains to be sacred. For the northernmost Tewa (San Juan, Santa Clara, San Ildefonso) the cardinal mountain of the east is Stone Man Mountain, 13,102-foot-high South Truchas Peak. For Nambe and Tesuque, farther south, the sacred mountain of the east is Blue Stone Mountain, or Lake Peak, from which rise the headwaters of the creeks that sustain both pueblos. Still other mountains of the Sangres are held sacred by the Tiwa of Taos and Picuris, including Jicarita Peak, which is located hard up against the northern boundary of the Pecos Wilderness. In 1916 ethnographer J. P. Harrington reported that members of Picuris religious fraternities visited the Jicarita summit frequently.

One of the most important functions of the sacred mountains is to mark the edges of the Pueblo homeland. Within their perimeter the work of creation is considered to be fully accomplished, and the Pueblo gods rule with complete control. But outside the ring of sacred mountains the world is less

Photo 3. Shrine on summit of Chicoma Peak. Photo by Edward S. Curtis, c. 1904. From The North American Indian, *courtesy Rare Book Collection, University of North Carolina at Chapel Hill.*

seh t'a, less finished, less ripe. The land beyond the mountains does not belong to the orderly part of the cosmos. Like the land that the four pairs of brothers explored, it is still soft and hazy, at least in a spiritual sense, and the protective umbrella of home deities is incomplete. Men have always gone to those regions from the pueblos, for instance to hunt bison on the Plains, but in going they took care, through ritual or fasting or the practice of sexual continence, to store enough spiritual power to last them through their journey and to compensate for the power that would be lacking in the distant land. To be outside the sacred mountains was to be off the spiritual map of the world. It was to be, in the oldest, most meaningful sense of the word, in wilderness.

The mountains are active forces in the spiritual landscape of the Pueblos. Elsie Clews Parsons even suggests in *Pueblo Indian Religion* that the mountains were no less than deities themselves, since they were represented in effigy in numerous rituals that take place within the pueblo. Whatever rank one may ascribe to them, the sacredness of the mountains signifies not the rarification of the life force but the ubiquity and abundance of it.

The mountains are the cloud gatherers, the rainmakers, the targets of the lightning bolts, and the Pueblos seek to control their power by means of a summit shrine called a *nan sipu*, which is yet another kind of earth-navel. The shrine is a key-hole arrangement of stones: a circle with a narrow, funnel-shaped opening that points in the direction of the pueblo. The *nan sipu* is held to be one of the few places in the Pueblo world where the separate subworlds of the Below, the Middle, and the Above come in contact with one another, and where, as a result, the mind of the spirit world is most accessible to the minds of men. The purpose of the *nan sipu* is to collect the blessings of all three of those subworlds and to beam them— the analogy to a radio transmitter is not far off the mark—to the center of the pueblo where the mother-earth navel receives and redirects them among the homes and fields of the people. In this way, all that is good in the Pueblo world is continually circulating through it.

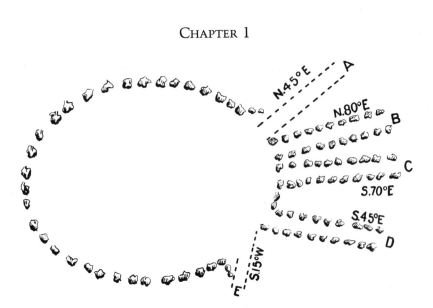

Photo 4. Diagram of Chicoma nan sipu *by William B. Douglass, who hypothesized that entrance trail A pointed to Taos; B to San Juan; C to Santa Clara; D to San Ildefonso; E to Jemez. Douglass argued that a trail to Cochiti should come between D and E. Courtesy Duke University Library.*

The correspondence between the physical landscape and the spiritual landscape of the Pueblos is further underscored by the Pueblos' use of the mountains as a pattern for architectural design and harmony. Either consciously or unconsciously, they have succeeded in building structures that resonate visually with the blue, distant mountains. The long elegant adobe church at San Ildefonso, for instance, seems to repeat the outline of sacred Chicoma and the rest of the Jemez Range, which looms behind it. The same kind of repetition occurs at Tesuque, where the roofline along the east side of the main plaza mimics the high, close horizon of Lake Peak and its ridges. Again and again at other pueblos the old buildings seem to silhouette the mountains and mesas, but nowhere is the influence of the landscape more strongly felt than at Taos, where each of the two great pueblo buildings *is* a mountain, like the mountains behind them.

It is easy to make too much of this kind of thing—architectural repetitions of the horizon begin to suggest themselves at every turn, like faces in the clouds—but the fact remains that the old buildings of the pueblos harmonize remarkably well with the landscapes around them. Unlike the modern, federally financed, and very comfortable "ranch" houses that now proliferate at the pueblos, the old, worn structures around the dance plazas seem not so much to have been built as to have occurred, like landforms.

In *The Sacred and the Profane*, Mircea Eliade explains that one enters sacred space in order to gain insight into reality, in order, that is, to receive religious experience. By extension, any place where such experience occurs acquires sacredness,

Photo 5. Taos Pueblo by John K. Hillers, 1880. Courtesy Museum of New Mexico.

at least in a personal sense. All of us have sacred places—our childhood homes or the places where we first fell in love or risked grave danger—and we complement them with innumerable sacred texts as well—not just Bibles, Talmuds, and Korans, but all the other things we read with a mind to asking and answering basic questions of life. While the Pueblos have sacred oral "texts," it is nonetheless not hyperbole to say that the sacred spaces of their landscape are also a text that they "read" or experience in a variety of ways in order to grapple with the fundamental problems of being and becoming. In the immediate vicinity of a pueblo there may be any number of small shrines—here a pile of stones associated with hunting small game, there a rock outcrop where one seeks spiritual aid in so mundane an occupation as cutting the leather soles of moccasins. And scattered through the surrounding hills and mountains are innumerable other shrines, some consisting of no more than an oddly shaped stone set on end.

One of the most important sacred spaces in the Pueblo homeland is the lake through which the First People emerged into the present world. For the Tewa that lake lies far to the north, beyond San Antonio Peak, in what is now Great Sand Dunes National Monument. In spite of the distance, the Tewa remain connected to the power of that place since each of their sacred mountains is associated with a lake or pond that replicates the original lake of emergence. The lakes are full of the spirits of those who "never did become"—the godlike beings who remained in the Below even after the Pueblo people emerged and who never tasted the air of the world where the sun shines. The spirits of certain mortal men dwell in the lakes as well. In life these men—and women—were "Made People," the highest order of Pueblo society. It was they who as caciques, medicine men, and ceremonial leaders guarded the secrets of Pueblo religion and guided their people through the complex rituals that ensured the pueblo's survival and well-being. In death the spirits of these leaders flew to the mountains and entered the lakes.

Years ago the sacred mountains and their lakes were visited regularly for retreat, for ceremony, and for prayer. In the early

1920s the noted photographer and ethnographer Edward S. Curtis wrote that the Tewa of Nambe conducted initiation ceremonies involving immersion and ritual bathing at Nambe Lake on Lake Peak, or alternatively at "a lake on Baldy peak a few miles north of Lake peak"—probably Spirit Lake, whose name derives from its reputation as a site for Indian ceremony. Curtis noted that the Indians commenced their ritual at dawn "so that no Mexican may observe them." A few years later Edgar Hewett, another early Southwestern ethnographer, reported that Lake Peak had special ritual significance with regard to fertility. He mentioned, with a discrete lack of elaboration, that the "mating ceremonies" of several pueblos were held annually at Nambe Lake in Springtime.

Powerful places, those lakes, especially Blue Lake in the mountains above Taos Pueblo. When the Forest Service appropriated Blue Lake and much of the surrounding mountain country as part of Carson National Forest in 1906, Taos Pueblo all but went to war. Blue Lake was its most sacred shrine and the site of some of its most important yearly rituals. Although the Forest Service and the Pueblo worked out a marginally acceptable *modus vivendi* for a time, the question of ownership came to a head in the early 1960s when the lake and its basin attracted the interest of increasing numbers of loggers and recreationists. The Taos pressed their case in the courts, in Congress, and in the Departments of Agriculture and Interior, and many non-Indians in New Mexico, arguing for the preservation of cultural and religious diversity, rallied to their cause. Finally in 1970 the lake and 48,000 acres of surrounding mountain forests were returned to the Pueblo. The celebration that followed was the celebration of a people who had survived a cultural apocalypse.

The Pueblos say that the power of the lakes makes them dangerous for the uninitiated. Under certain circumstances only Made People may safely go to them, and even they do not take their journey lightly. Once a Taos Indian, speaking of Blue Lake, explained to me that "there is a kiva beneath the surface of the water, and it goes down and down without stopping. If you let it, it can pull you in, and you will never come out." Blue Lake is strictly off limits to backpackers, and patrolmen guard it vigilantly; perhaps its power is as strong as it ever was. One wonders, though, about the lakes associated with other more public mountains like Lake Peak and Truchas Peak in the Pecos Wilderness. One must expect that the intrusions of herdsmen and hikers have weakened their spiritual authority. Certainly many rituals have had to be relocated, if not abandoned.

People who read and write books find it natural to refer to the "physical" landscape of the northern Rio Grande valley and to say that the Pueblos superimposed on it a "cosmic" or "spiritual" landscape of sacred sites, of ethereal blessings flowing from place to place, of landforms and locations considered animate, or nearly so. This view, however, beggars what the Pueblos themselves traditionally believed. To them the world was not two things spiritual and physical joined together; it was a single, indivisible whole. Spirituality was an essential quality of all creation, and the Taos who venerate Blue Lake or the Tewa who arranged the *nan sipu* atop Chicoma would have found it difficult, if not impossible, to conceptualize so inanimate a thing as a merely "physical" landscape. In this they would certainly not be alone. The Navajo and the Jicarilla Apache, who were the Pueblos' two most permanent neighbors, found in the landscape an equally rich spiritual resonance.

The Pueblos explain the existence of these non-Pueblos in more or less the same way that anthropologists account for human diversity—in terms of physical separation and subsequent differential development. They say that after emergence from the Below all people were one until they came to a wide river, where Magpie, one of the spirits accompanying them, stretched out his long tail to make a bridge. A portion of the people crossed over, but suddenly and capriciously Magpie lowered his tail into the water. The people who were swept into the current instantly became fish, and with no bridge, the people on one side of the river were now irrevocably separated from those on the other. As time went on, each group developed different identities. Those who had crossed were the Pueblos; those left behind became Navajo, Ute, Apache, Kiowa, and Comanche.

From an anthropological point of view, the river that separated the Pueblos from the others lay far from New Mexico. The Navajo and Apache are Athapaskans who probably developed in west-central Canada near where their linguistic relatives, the northern Athapaskans, still live. No more than a thousand years ago they left their home in the far north and began to migrate southward. The Jicarilla arrived in New Mexico about 1500 by way of the eastern plains and the foothills of the Rockies. The Navajo and the western Apaches, meanwhile, probably followed an intermountain route farther west and reached the Southwest a little later.

The Navajo settled in the Four Corners country where the Anasazi had lived before them, and like the Pueblos, they defined their homeland in terms of sacred mountains: in the south, Mount Taylor, near Grants, New Mexico; in the west, the San Francisco Peaks, above Flagstaff; in the north, a peak in the La Plata Range of southwestern Colorado; and in the east, Cabezon Mountain, west of Jemez Pueblo. Today, as proprietors of a reservation the size of West Virginia, the Navajo still occupy much of their traditional territory.

Originally the Southwestern homeland of the Jicarilla comprised an area east of the Pueblos comparable to the territory of the Navajo to the west. Pressure from other plains tribes, principally the Comanche, however, forced the Jicarilla to move into and ultimately across the Sangre de Cristo Range so that their territory overlapped that of the Pueblos. In the Jicarillas' world view, the earth was a woman lying on her back, and the sky a man lying on top of her. Humans were their children. Although since 1887 their only land has been a reservation along the continental divide in northern New Mexico, the Jicarillas traditionally conceived their homeland to include most of the upper watersheds of their four sacred rivers—the Pecos, Canadian, Rio Grande, and Arkansas (for which the Chama was sometimes substituted). The Pecos and Canadian were considered female rivers, the others male. Since an important rite for every newborn child required water from a "married couple" of one male and one female river, the Jicarilla must have been accustomed to traveling fairly continuously over their broad territory.

The Jicarilla believed that the Sangre de Cristo Range was a spine running through the center of their homeland and that the heart of the earth was a spot very close to Taos Pueblo, a notion with which the people of Taos would undoubtedly agree. The Jicarillas ranged over the mountain country, visiting often at Taos, San Juan, Picuris, and, in its days of prosperity, Pecos. They trekked yearly to the southeastern slopes of the Sangres to bathe at the hot springs of Gallinas Canyon, at least until the Atchison, Topeka, and Santa Fe Railroad developed the springs as a tourist resort.

The Jicarillas' relation to the land was different in important respects from that of the Pueblos. Although the Jicarillas practiced agriculture on a limited basis—and would have done so more had circumstances of war and peace permitted—they subsisted primarily by hunting and gathering. They were less concerned than the Pueblos with the power of sacred sites to bring rain or otherwise control the weather. Instead they looked to the mountains as a source of game and other food and also as a place to find refuge from their enemies. In this respect they had much in common with the earliest known mountaineers of the southern Sangres, men and women who lived at about the time of Christ or somewhat before and whose passage through the mountains we may still detect today.

2

Into the Mountains

There is a spirit of energy in mountains, and they impart it to all who approach them.

— *Francis Parkman*

THE EARLIEST PEOPLE to frequent the high Sangres remain a mystery in many ways, in part because archaeologists have given them scant attention until relatively recently. We are not sure exactly who they were, what they sought in the mountains, why they made their camps where they did, or even what to call them.

They were gatherers and hunters whose closest cultural contacts lay with the Oshara tradition of the Desert Culture, centered to the west. Their living arrangements were flexible and mobile. They had few possessions, built most of their shelters to last only weeks or months, and irregularly congregated in groups of several dozen or dispersed in small family bands. Notwithstanding their seasonal wanderings, they were also part-time agriculturalists and planted small patches of maize, squash, and possibly beans in canyon bottoms scattered throughout their territory. The more plots and the more different soils and climatic niches they sowed, the more they protected themselves against drought, flood, and other hazards including theft. Even so, their harvests were rarely abundant, for these early farmers probably abandoned their crops to the

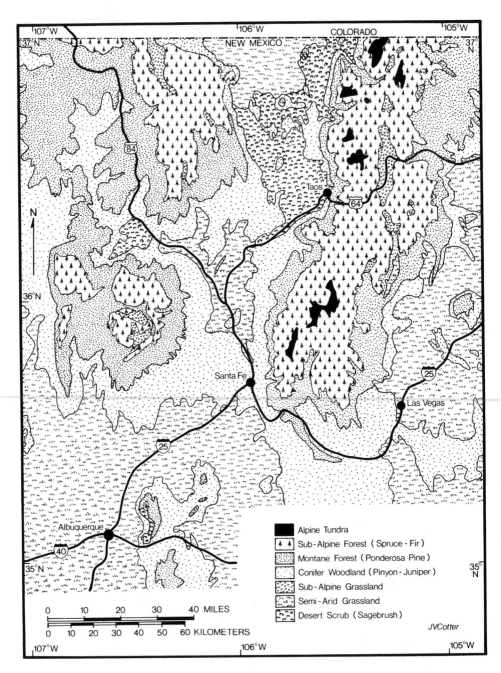

Map 3. Vegetation of north central New Mexico.

depredations of rabbits, deer, and the like, returning only at summer's end to harvest them as they would a crop of wild berries or roots.

In hunting, as in agriculture, the early people minimized their risk of failure by exploiting as wide as possible a range of environments. Presumably that is why they climbed to the highest reaches of the mountains, and although they may not have been the first human beings to do so, they were the first to leave a trail there that we in the present can follow. We guess, but cannot be sure, that these highly mobile people climbed into the mountains in pursuit of the meat and hides of the larger animals: deer, elk, bighorn, and possibly bear. Although in that era (and for centuries after) the big-game species were not restricted to high altitudes as they are now, large numbers of them nonetheless followed the receding snowline into the high valleys when summer grasses grew deep and lush. Band by band they filtered upward through the forests, and the hunters followed them into the heights.

In ascending the mountains, the hunters journeyed through a variety of ecological zones, each one successively colder and wetter than the one before, and each one characterized by its own unique assemblage of plants. From the semidesert shrub and grasslands of the Rio Grande valley they climbed into a woodland of piñon pine and juniper, where the tree canopies grew round and wide in order to shade their roots from the parching sun. From a distance the widely spaced, slow-growing trees looked like a green pox on the bare brown skin of the soil, and in autumn in this ecological zone the hunters devoted several weeks to collecting the nutritious nuts of the piñon.

As the hunters climbed and the woodland grew thick and shady, ponderosa pine began to replace the piñon and juniper. Tall and heavy-limbed, the older ponderosa trees had thick orange bark that years of steady growth had split into large plates. The cracks between the plates smelled of vanilla; when the wind was still the forest was redolent with it. Wild turkey roosted in the tall pines, and in the forest understory grew thickets of Gambel oak, whose acorns fed bear, squirrel, and human alike.

The soil became markedly darker and richer in organic materials as the ponderosa very gradually gave way to a mixture of taller, skinnier conifers: limber pine and white fir on dry sites, and blue spruce in the wet soils of the canyon bottoms. The most common tree was the Douglas fir, which is the tree the Tewa say they climbed in order to emerge from Sandy Place Lake into the present world. From the clear and open country of the valley the hunters had come now to the heart of the mountain forest. The land was rugged and enclosed, tangled with dead and wind-felled trees and almost invariably steep. As they continued to climb into the increasingly cool, thin air of the higher slopes, the Douglas fir gave way to Englemann spruce and alpine fir, which grew close together on the now shallow, stony soil, their roots tightly intertwined and their tops crowded into a single unbroken canopy. The understory of the spruce-fir forest was so dim that few plants grew there; hence there was little feed and little reason for large animals to linger in the subalpine forest. The hunters understood, however, that they could open the forest to sunlight and enhance their prospects for hunting by making judicious use of their single most powerful tool: fire.

Primitive people throughout the world have used fire to change and influence their environment. Usually their purpose has been to convert forest to grassland or to protect grassland from the encroachment of trees. The growing literature on this subject makes clear that early man's use of fire was not usually haphazard but instead prudent and thoughtfully grounded in an extensive understanding of the effect of fire on various types of vegetation.

In the southern Sangres early hunters probably used fire to create and maintain forest openings and even to burn off entire mountain sides. Certainly natural, lightning-caused fires also accomplished these ends, but since optimal fire conditions occurred irregularly in the wet, cool environment of the high country, it is unlikely that early hunters relied entirely on chance. On slopes with relatively warm exposures their fires helped clear the way for spruce and fir to be replaced first by shrubs and grasses and later by aspen, a white-barked, deciduous poplar whose leaves turn burning gold with the frosts of fall. Since the canopy of the aspen forest is leafless during the early part of the growing season and still relatively open during the remainder, grasses, forbs, and the animals that depend upon them prosper beneath it.

With the exception of a few timberline groves of bristlecone pine, no trees in the southern Sangres grow at altitudes as great as Englemann spruce and alpine fir. Around the highest peaks and divides they reach their environmental limits for cold and wind, and the trees become thin, somewhat stunted. The forest gradually thins out to an open parklike wood and then yields to a tawny alpine grassland dotted with clumps of nearly prostrate, grotesquely formed spruces called krummholz or wind timber. In these most harsh and wintry reaches of the mountains the protection from the wind that a group of trees provides to its individual members is essential to their

Photo 6. Wind timber or krummholz at about 11,800 feet on Trailriders Wall, which is the high divide linking Pecos Baldy to the Truchas Peaks. Author photo.

Photo 7. Truchas Peaks from Trailriders Wall. Early hunters and gatherers camped in the low saddles of this divide. Author photo.

survival. In fact, whole islands of krummholz actually move across the alpine tundra in a slow wind-driven migration, as their windward sides continually die back and the leeward sides grow and reproduce.

The bald alpine zone above timberline is almost a desert. In winter the snow that falls on it is whisked off by the wind. Its soil is so shallow and its slopes so steep that rainwater quickly evaporates or drains away. Although it receives four or five times as much precipitation as the near-desert in the valley, its plants tend toward succulence, like the plants down there. Many of them, including the various alpine willows that form a mat on the highest peaks, have fleshy leaves with thick skins to conserve water. The alpine plants also favor reddish colors in order to screen out the harsh ultraviolet radiation that at high altitudes is especially intense. They are, in short, residents of an exceptionally cruel environment. In all the Southwest there is no other place that so combines cold, drought, and radiation with the violence of storm and wind. Yet, incredibly, it was in these spaces that the early hunters often made their camps.

Nearly all the prehistoric encampments that have been identified in the high country have been located in the cold windy saddles of the high divides, well above timberline, where the climate is rarely benign, where hailstorms and freezing temperatures are common even in August, and where water in some cases is as far as a mile away. The only explantion commonly offered for the placement of the camps is that they afforded a good vantage from which to watch for game animals. But this is really no explanation at all. Hunters and game managers today frequently climb the divides in order to watch for wildlife, but they do not camp on them. They camp down in the sheltering cirques, where water is available. The cirques should have been doubly attractive to the early hunters since a number of edible wild plants, like the marsh marigold and alpine thistle, grow abundantly in them. The hunter bands could have as easily camped there and posted lookouts on the heights during the day. Admittedly, one cannot say for sure that the hunters did *not* camp in the cirques. Although no such camps have been found, they may simply be hidden by the build-up of organic matter on top of them.

Conversely, there is little on the high ridges to conceal the remains of an encampment, even one that is two or three thousand years old. These remains consist primarily of rings of stone that formed hearths, of blades and scrapers and grinding stones, of all sizes of projectile points, of workshop debris, and of a miscellany of small stones foreign to the mountains. By comparison with similar assemblages from other archaeological sites in New Mexico, the majority of the high-country materials have been dated between fifteen hundred and three thousand years old, with the greatest probability that they were manufactured at about the time of Christ or slightly before. This raises the intriguing possibility that, except for the last two centuries, the most intensive human use of the high country occurred in the last few hundred years B.C.

But why? It is frequently assumed that people exploit the more inaccessible mountain environments because of population pressure at lower elevations. The northern Rio Grande valley at the time in question, however, is generally agreed to have been sparsely settled. Certainly its population was much less than it would be a few centuries later, after use of the high mountain country had virtually ceased. And if population were low, then one would expect the people of the valley to respond to periods of drought and scarcity by migrating to environments more convenient than the alpine summits. Perhaps, then, they came to the mountains, not driven by dire issues of survival, but to obtain certain preferred animal, plant, or lithic resources that were not available elsewhere. Whatever the reason, the Anasazi who succeeded them would not find it compelling.

It is generally agreed that the people of the Oshara tradition, to which the gatherers and hunters of the Sangres belonged, were the eventual progenitors of Anasazi culture. Nevertheless, there is nothing to indicate whether the descendants of the men and women of the high mountain camps actually contributed to Anasazi and Pueblo cultural development or were driven off and replaced in the Rio Grande valley by other Oshara tradition groups who did. In any case, by about A.D. 600 life in the valley began to change dramatically. The bow and arrow came into wide use, as did pottery, which revolutionized food preparation and storage and also opened rich new possibilities for artistic expression. Not much later, new and higher-yielding varieties of maize appeared in northern New Mexico, making agriculture much more reliable and successful.

Beginning about A.D. 900 the population along the Rio Grande began to grow rapidly, but not nearly so rapidly as it did in the canyons and mesas to the northwest. As Anasazi culture in the Four Corners country waxed and waned through its "Golden Age" of 1050 to 1250, the people of the Rio Grande, who were probably the progenitors of today's Tiwa, lived in comparative simplicity and isolation. It is not at all clear to what extent the Tewa, Towa, Tano, and Keres may also have been indigenous to the valley, but their presence is well documented following the general abandonment of the Four Corners from about 1300 onward.

The immigrants to the Rio Grande were obliged to blend or fight their way into the existing pattern of settlement, and as population and probably conflict in the valley increased, people continued to move eastward past the mountains to the Rio Pecos and into the upper reaches of the Canadian River watershed. The migration to the edges of the plains may also have been spurred in part by a simultaneous expansion of the range of the bison and a correspondingly greater emphasis on hunting among these easternmost Pueblos.

As the situation began to stabilize, new centers of population began to emerge at Puye (a few miles west of Santa Clara), Tyuonyi (the cliff dwellings of Bandelier National Monument), Taos, and elsewhere. Several hundred Towa-speaking people congregated in a village at a place now called Forked Lightning Ruin, a short distance downstream from the future Pecos Pueblo. Seventy miles to the north a small mountain community on the banks of the Rio Pueblo suddenly swelled with growth, laying the foundations of what would become the Pueblo of Picuris. And along the western edge of the mountains numerous small pueblos sprang up, including the ancestral homes of the Tewa.

It may be foolhardy to hypothesize from an absence of data, but apparently the people of the upper Rio Grande made slight, if any, use of the high mountains for about a thousand years from A.D. 500 onward, even though they congregated in numerous villages in the foothills. One wonders why. Did a generally elevated level of warfare in the region make prolonged summer absences dangerous and undesirable? Did the innovation of bean cultivation keep farmers closer to their fields, and did the new advantage of the bow and arrow enable them to satisfy their families' requirements for meat without ranging so far, or for so long, from home? As Puebloan people began to colonize the Plains after about A.D. 1100, they learned to exploit a source of meat superior to anything the mountains had to offer: the wild bison, which, weighing more than a ton, was three times the size of an average elk.

By about 1300 the Pueblos of the Rio Grande made regular expeditions to hunt the bison on the eastern plains or to trade for its dried meat at the eastern Pueblos. As long as access to the plains was free, there may have been little need to hunt extensively in the mountains, but beginning about 1500 the Pueblos encountered a new breed of horseless bison hunter who appeared from the north. These Athapaskan nomads, probably including the ancestors of the Jicarilla, rendered the Pueblos' expeditions to the plains increasingly dangerous, and their raids on Pueblo villages forced the Pueblos to band together into ever larger and more defensible communities. Eventually, rather than hunt bison, the Pueblos traded for it with corn and other agricultural products. The Apaches became their partners during times of plenty and their enemies when food was short.

It may not simply be coincidence that evidence of human use of the high country picks up again with the entry of the Apache. Potsherds of either Galisteo (an extinct Tano pueblo) or Picuris origin have been found in a meadow in the gap between the Truchas and Santa Fe ranges, and an intact Picuris pot was discovered in a clearing near the Rito del Padre, one of the headwater streams of the Rio Pecos. Both finds were dated to about 1500, and they indicate the presence of either Pueblos in the high country, or just as likely, Apaches who had obtained the pots by trade.

After 1500, human circumstances changed ever more radically and rapidly in the land beneath the mountains. A long period of continuity was at its close. Although different in many ways, the Anasazi, Pueblos, and Apaches were united in the broad outlines of their views of life and in their limited capacity to manipulate the world around them. They understood time to be a continuous, ahistorical present and space a spiritually resonant and animate landscape. Into this world marched the Spaniard Coronado and later Oñate and his colonists, bringing metal tools and weapons, gunpowder, wheeled vehicles, livestock (which would change the Southwestern landscape as nothing else had ever done), and myriad other new material things that vastly increased man's leverage on his environment and on his fellow man. The newcomers believed in a historical, nonrepetitive time that progressed according to divine plan toward a day of universal judgment. They believed they executed God's will by creating a rigid, central government backed by military force and superior to the authority of any pueblo. It was their mission to exercise dominion over the land and native peoples of New Mexico for the sake of His greater glory and, if possible, their own material gain. In time many of the Spaniards' attitudes would become more like those of the Pueblo, but at first the two groups shared little but their penchant for ceremony and their dependence on agriculture. It would take them more than a century to learn how to live together. The pages to come describe how two pueblos, Pecos and Picuris, responded to the threats posed by the Spanish and how finally they joined with them to face yet another threat from beyond their frontiers—the Comanche of the southern plains. Although centuries of conflict weakened the Pueblos tragically and threatened their very existence, their great achievement is that they survived as a people, and also, more remarkably, as a culture.

3

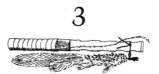

Brothers on the Faultline

*It is pitiful to view this immense area and the
many ignorant people who inhabit those vast
regions, all without knowledge of the Blood of
Christ or of His Holy Faith.*
> —*Gaspar Pérez de Villagrá, 1610*

A MAJOR, THOUGH LONG DORMANT geologic fault runs through
the high country from Pecos in the south as far north as Picuris. Arrow-straight, it cracks the mountain bedrock for more
than thirty-eight miles, and in the light of history, the fact
that the fault connects those two particular places makes a
bitter irony. During the seventeenth and eighteenth centuries,
as both pueblos were pushed to the edge of oblivion, many of
the people of both Pecos and Picuris must have felt that the
world had indeed been rent by a giant crack and that they
were falling into it.

The fate of the two pueblos was tied to the mountains. Each
occupied a favored geographic position, a gap through the
mountains that placed it at the nexus of the wild, buffalo-
filled plains and the settled, agricultural Rio Grande valley.
There were few such gaps, and so the pueblos that controlled
them were natural trading centers. Pecos, especially, grew rich
from the lively exchange of Pueblo pottery and corn for the
hides, meat, and tallow brought in by Plains Apaches. The
double exposure of the frontier pueblos, however, could turn
to double liability as the pressures of both worlds bore down

upon them. Caught between the raiding of plains nomads and the harsh hand of Spanish rule, the frontier pueblos struggled to survive, and they did not always succeed.

The balance they had achieved with their environment prior to the arrival of the Spanish now slipped out of their control. Spanish horses in the possession of the plains tribes gave the pueblos' enemies a tactical advantage. European diseases, against which they were biologically defenseless, decimated their populations. During the seventeenth century the Tompiro pueblos, which lay south of the Sangres on the actual plains, perished entirely; Galisteo, also on the frontier, was abandoned twice during the eighteenth century, the last time for good; and Pecos, once mighty, was reduced to a band of seventeen refugees who trudged away from the ruined pueblo in 1838, never to return. Picuris, which also was abandoned for a time, barely survived the years of tumult, and of all the pueblos of the eastern frontier only Taos, which also guarded a mountain pass of importance, endured the first two centuries of European occupation with anything like its original strength and vitality.

At the time of the Spanish entrada in 1540 Pecos and Picuris were unsurpassed in the Pueblo world for their power and size. Their buildings were four and five stories tall, as Taos is today; each boasted a population of nearly two thousand. Picuris lay in a mountain valley at the junction of two rivers, one of which led to a notch in the mountain divide that overlooked the splendid Mora Valley and the plains beyond. This pass, like the one at Taos, was steep, rugged, and often impassable because of winter snows, yet it afforded the Picuris and their trading partners a freedom of movement that most other pueblos lacked. The easiest passage to the plains, however, and that which connected most directly to the Spanish capital at Santa Fe, lay through Pecos. This pueblo perched atop a small mesa between the foot of the Sangres and the imposing, square-edged bulk of Glorieta Mesa. That the location was a strategic one has been attested by American roads as well as Spanish and Indian ones. The Santa Fe Trail passed within sight of the pueblo, and its ruts may still be traced

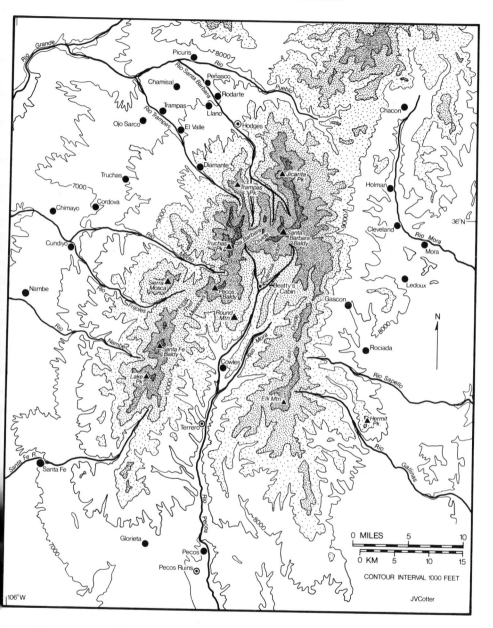

Map 4. Landforms and settlements of the southern Sangre de Cristo Mountains.

across the small plain where trading parties of Apaches used to camp. The first railroad to enter New Mexico passed by the pueblo too, near enough that its proprietors erected a billboard to promote the by-then abandoned ruins as a tourist attraction.

Today there is little at either Picuris, a struggling community of slightly over a hundred people, or at Pecos, now only a ruin in the care of the National Park Service, to suggest how vital and grand those places once were. But three centuries ago the affairs of the two pueblos preoccupied New Mexico's Spanish conquerors—Picuris because it was a hotbed of rebellion and discontent; Pecos because it was the sometimes restless ally without which the Spanish experiment would have failed.

In 1540 Francisco Vásquez de Coronado planted his banner in New Mexican soil, claiming the entirety of the new land in the names of his Spanish King, his Catholic Pope, and his Christian God, all of whom belonged to a space and time quite outside the Pueblo cosmos. The fate of Coronado and his small army is a dramatic lesson in the limits to the power of positive thinking. So convinced were they that great wealth lay in the northern wilderness that even after the fabled Seven Cities of Cibola turned out to be the six mud villages of Zuni, and even after the villages of the Rio Grande valley proved to be equally destitute, Coronado and his men continued to chase their illusions out across the buffalo plains to the squalid grass huts of the Wichita Indians in central Kansas. The image of these fierce soldiers of cross and crown riding into the treeless plains, accepting the counsel of treacherous guides, and possessing no maps or proof of any kind that the object of their search existed, except of course, for such evidence as they invented to convince themselves—it is an image of folly surpassing anything Don Quixote ever fantasized, and yet it is also an image full of serious meaning for the history of New Mexico. Avarice may account for the ultimate failure of the Coronado expedition, but avarice alone cannot explain its achievements. Coronado and his followers were proud men of a proud civilization, and they approached the task of conquest with a sense of national and religious mission. As much as anything, it was their abiding faith in their own greatness that sustained them in their extraordinary exploits, and many of those who followed them to New Mexico would possess that faith in equally large measure.

Although Coronado found no golden cities, and the half-dozen or so expeditions, authorized and unauthorized, that followed him ended in similar failure, the simple fact that the natives of the region lived in towns, even small and labyrinthine ones, indicated to the Spanish that the northern land possessed the potential both for wealth and for the spread of Spanish civilization. Quite unlike the half-naked Chichimecs who harassed Mexico's northern frontier, the Pueblos were *gente vestida*, clothed people:

most if not all the men wore cotton blankets and on top of these a buffalo hide. Some covered their privy parts with small cloths, very elegant and finely worked. The women wore . . . turkey-feather cloaks and many other novel things—all of which for barbarians is remarkable.

The existence of these remarkable barbarians took on special importance as the focus of the Spanish frontier shifted from conquest to mineral development. Finding no great cities north of the former Aztec empire, the Spanish sought riches instead in the ores of Mexico's mountains and undertook to extract them by means of Indian labor. Silver-rich Zacatecas, in 1548, became the first town of the new mining frontier, and others including Guanajuato and San Luís Potosí soon followed. Development of the mines was hampered, however, by the intractability of the wild Chichimec Indians. The more settled Indians of New Mexico, by contrast, promised greater productivity, and the abundance of mountains in the far north seemed a guarantee that plenty of ore might be found.

One of those who thought so was Don Juan de Oñate, who grew up on the northern Mexican frontier and was thoroughly familiar with mining, and also with the associated industries of slaving and stock raising. In hopes of building a lucrative new frontier in New Mexico he brought the first permanent Spanish colony north at his own expense in 1598 and settled it at the confluence of the Chama and the Rio Grande. Unfortunately, however, Oñate's colonists found no mines nor riches of any kind. Instead they staved off starvation only by seizing the food stores of their Tewa neighbors. Within six months of the founding of the colony, rebellion broke out at Acoma, and Oñate responded to it with the utmost savagery, slaughtering hundreds of Acoma people and imposing on the survivors a punishment of amputation and slavery. The affairs of Spain's new northern outpost could not have begun less auspiciously, and the resolve of the colonists began to weaken.

Conditions grew so bad that many of the colonists talked of desertion, which is what, in fact, their unauthorized return to Mexico would have been. Unlike pioneers on the Anglo-American frontier, the Spanish colonists were not free to withdraw. By law they owed Oñate the same obedience a soldier owes his general. But they were hungry, their hopes for riches and glory were dashed, and they had no prospects, save that the undisguised hatred of the Indians promised further bloodshed. In 1601 while Oñate hunted treasure on the buffalo plains, a majority of the colonists fled down the Rio Grande toward home.

Following the mass exodus of the colonists, the effort to colonize New Mexico probably should have been abandoned, and it would have been except that the Franciscan missionaries Oñate had taken to New Mexico were loath to give up so immense a territory which they alone held the right to evangelize. Exaggerating their actual accomplishments at least tenfold, the Franciscans claimed to have baptized more than seven thousand heathens among the Pueblos. They entreated the Crown not to turn its back on these converts, and the Crown granted their wish. His majesty Philip III of Spain ordered that the colonization of New Mexico continue, even at royal expense. Oñate was recalled and his proprietary charter revoked. Officious Don Pedro de Peralta rode north to assume the newly royal governorship of the foundering colony.

Peralta made his headquarters a respectful distance from the powerful Tewa pueblos of the Española Valley in a more peaceful and defensible location hard against the foot of the

Photo 8. Mission, Picuris Pueblo. Photo by Adam Clark Vroman, 1899. Courtesy Museum of New Mexico.

Sangres. Here he founded a new capital for New Mexico, *la villa de Santa Fe.* The Franciscans, meanwhile, established their headquarters twenty miles from the city of Holy Faith, at Santo Domingo Pueblo, and the physical separation of the two centers of authority was indicative of the grave divergence of affection and purpose that existed between the Two Majesties, the Pope and the King of Spain, who jointly ruled New Mexico. For seventy years, until the successful rebellion of the Pueblos in 1680, the Franciscans and the colony's civil authorities struggled incessantly with each other. Peralta himself was excommunicated twice and ultimately imprisoned for his opposition to the friars.

The conflict between church and state, which frequently verged on civil war, had its source in the colony's poverty. New Mexico possessed only one resource of any value, and that was the energy of its native people. Both the Franciscans, whose lofty aspirations included the building of innumerable churches, chapels, and other mission facilities, and the various governors and their lieutenants, who had come to New Mexico to enrich themselves, desired to exploit it. The resultant burden of work on the Pueblos was immense, as the experience of Pecos Pueblo attests. Between 1621 and 1625 the natives there were compelled to build a monumental fortress of a church, 40 feet wide, 40 feet high, and 145 feet long. Its walls, which were reinforced with massive buttresses, stood on 9-foot-wide foundations and consumed an astronomical 300,000 adobe bricks, all of which were made by the beleaguered natives. When finished, the church was a magnificent, if bizarre, architectural achievement: "a sixteenth century Mexican fortress-church in the medieval tradition, rendered in adobe at the ends of the earth."

In spite of their exhaustion from building the church, the Pecos were also obliged to pay exorbitant tribute to the civil authorities. Twice a year a valuable haul of hides, tanned skins, piñon nuts, and native cotton weavings was extorted from the natives, much of it later shipped south to Chihuahua, where it sold for a tidy sum, nearly all profit. The natives scarcely had time to plant and cultivate their corn, but in addition to the economic hardships imposed by the Spanish, the Pecos and other Pueblos also had to bear the repression of their traditional religion. In the Pueblo worldview there was no heaven and no hell. Life was sanctioned and understood as focusing entirely on its present materialization—on community, on the weather, on crops, on successes of love, health, and war. If these un-Christian understandings were but dimly perceived by the friars, the feathers and dances and stone idols that went with them were not. With the same intolerance that had led their predecessors to demolish the temples of the Aztecs, the New Mexico Franciscans raided kivas, forbade ceremonial dances, destroyed sacred masks and fetishes, and required—under threat of corporal punishment—regular observance of the Christian sacraments.

The Franciscans were checked in their zeal only by the studied uncooperation of the Spanish governors of New Mexico and by the resistance of the Pueblos themselves. Picuris was especially troublesome. Because of sustained hostility to the Franciscan stationed at the pueblo, including several attempts on his life, the Picuris mission was abandoned for about four years beginning in 1624. At that time Fray Alonso de Benavides, the custos (superior) of the New Mexico Franciscans, vented his frustration with the Picuris by labeling them "the most indomitable and treacherous people of this whole kingdom." The Picuris, however, appear to have had some competition in that department, for in 1632 the Hopis killed the missionary stationed with them, and seven years later the Taos Indians killed theirs.

Full-scale revolution was a real possiblity in New Mexico throughout the seventeenth century, but while the Spanish representatives of church and state were tireless in blaming each other for the colony's instability, neither side undertook to find a peaceful solution. Instead the military forces of the colony kept the Pueblos in line by responding to rebellious behavior with swift and sanguinary punishment. Twenty-nine suspected rebels were hanged in Jemez Pueblo in 1647, and three years later, after an abortive revolt by the Jemez, Keres, and southern Tiwa, nine more incorrigibles were hanged and many others sold into slavery. The Spanish position in New Mexico grew still more precarious when a terrible drought descended on the colony in 1666 and lasted five years. The colony ran so short of food that the Spanish were reduced to roasting and eating the hides that were the principal furnishings of their simple homes. The Pueblos, of course, fared much worse; hundreds of them "perished of hunger, lying dead along the roads, in the ravines, and in their hovels."

The Pueblos interpreted the drought as a clear sign that the Spaniards' religion did not ensure the orderly and fruitful progression of the seasons as they believed theirs did. Furthermore, the Spaniards' promises of military protection proved to be worthless; outlying Navajo and Apache bands, who lately had acquired use of the horse and who suffered equally from drought and food shortage, stepped up their attacks against the colony.

At last in 1680 the Pueblos achieved the unity that had long eluded them, and they rose up together and drove the twenty-five hundred Spaniards of the colony back to old Mexico, killing some four hundred, including nearly all the Franciscans. Under the charismatic leadership of Popé, a San Juan medicine man, the Pueblos methodically destroyed every church and hacienda, every remnant of Spanish culture in New Mexico. The great church at Pecos was sacked and put to the torch, and then in an orgy of destructive energy, the colossal walls were thrown down brick by brick.

Aside from one abortive expedition in 1681, no effort was made to reconquer the lost territory for a dozen years. The Pueblos, meanwhile, remained loosely confederated, although old enmities gradually eroded the sense of brotherhood that

Photo 9. Pecos Mission ruins, September 3, 1880. These ruins are the remains of a church built in 1705 and subsequent years on the site of the much larger mission that the Pecos destroyed in 1680. Photo by George C. Bennett. Courtesy Museum of New Mexico.

the rebellion had produced. By the time Don Diego de Vargas headed north in 1692 to wrest New Mexico from Pueblo control, the rifts between the Pueblos were big enough for Vargas to march his army through.

At first virtually all the Pueblos submitted to Vargas peacefully, but their resistance increased as the process of recolonization proceeded. Vargas soon learned that the Pecos were his invaluable allies, as they repeatedly warned of plots against him and provided him with supplies and reinforcements. The Picuris, meanwhile, had returned to their "indomitable and treacherous" ways.

The Picuris rebelled in 1694, and avoided punishment by hiding in the mountains. They returned to their pueblo making peaceful protestations to the Spanish authorities, and then rebelled again in 1696, along with most of the other northern pueblos. Twenty-one colonists and five missionaries were slaughtered, but the revolt soon foundered for lack of support from Pecos and other pueblos. Vargas, at the head of an army that included many Indian auxiliaries, marched north to punish the rebels, and, expecting no mercy from him, the Picuris fled to the plains.

Vargas caught up with their rear guard somewhere east of the Mora Valley and succeeded in killing a number of men and capturing eighty-four women and children. The rest of the fugitive pueblo escaped, in company with Apache allies and friends, and Vargas returned to Santa Fe by way of Pecos through a driving, week-long blizzard. Once there, he divided the captives as slaves among the soldiers and colonists who had assisted him, noting that enslaving instead of killing them was an act of considerable compassion, considering "their desertion of our holy religion and their royal vasselage."

The fate of the Picuris who escaped Vargas's punitive expedition is less well known. Some of them gradually filtered back to their mountain home on the Rio Pueblo, but the vast majority continued north and east across the plains to a place called El Cuartelejo, where they settled with a tribe of Apaches who lived in scattered rancherias. Apparently some of the Picuris refugees were unhappy with their new situation, for word repeatedly reached Santa Fe over the ensuing years that the Cuartelejo Apaches were using them cruelly and they wished to come home. As a result, in 1706 *Sargento mayor* Juan de Ulibarrí, the chief official of the Pecos district, led an expedition to El Cuartelejo, which lay in either western Kansas or east-central Colorado, to ransom the Picuris and bring them back. Ulibarrí's expedition added greatly to Spanish knowledge of the lands and tribes beyond the northeastern frontier, but the number of Picuris who returned with him to Santa Fe was only sixty-two. Presumably many others either were held by Apache groups Ulibarrí did not contact or preferred life on the plains to life under Spanish rule and elected to remain with the Apaches.

Forty years earlier Fray Augustín de Vetancurt had reported three thousand persons living at Picuris, but now with the addition of the refugees, the population of the pueblos was still less than four hundred. Even allowing for considerable exaggeration by Vetancurt, the decline is astonishing. Certainly European diseases had taken their toll (although the worst epidemic of the seventeenth century, which killed more than a tenth of all the Pueblos, took place in 1640, before the Vetancurt count). Certainly too, the bloodshed of the 1680 rebellion had reduced numbers somewhat, and so had the hardship and violence of the flight to the plains, but none of the events of the period 1666–1706 seem adequate to account for so terrible a loss of population in terms of mortality alone.

Conditions in the Pueblo world during the late seventeenth century were so chaotic that several pueblos were utterly abandoned. The Tano inhabitants of Galisteo and San Marcos, for instance, moved en masse to well-watered valleys near Chimayo in the Santa Cruz Valley, although the Spanish later evicted them and forced them to resettle Galisteo. At Picuris the stimulus to leave was probably both political and environmental. Ever since Apaches arrived in the region, the Picuris had dealt closely with them, welcoming their presence. The Apaches' steadily increasing use of the mountains, combined with the demands of the pueblo's large population, probably overstrained the plant and animal resources of the area

Photo 10. Threshing with goats, Picuris Pueblo. Photo by
Ed Andrews, c. 1905. Courtesy Museum of New Mexico.

surrounding Picuris, significantly reducing the carrying capacity of the land. If this was indeed the case—and archaeological evidence indicates that the material wealth of the pueblo,
as measured by the size, number, and condition of its buildings, had already begun to decline by 1600—then many Picuris, given the added impetus of Spanish oppression, may
well have elected to seek new lives in the company of neighboring mountain Apaches and their plains cousins at Cuartelejo.

A further reason for the decline of the pueblo may have been
psychological. The long close association of Picuris and other
pueblos with the Apaches (and principally with the tribe the
Spanish would soon begin to call *Jicarilla*) had led the nomads
to adopt an increasingly sedentary way of life that featured
pottery making, irrigation agriculture, and semipermanent
adobe dwellings. At the same time, some of the individualism
and independence of the Apaches may have rubbed off on the

Picuris, eroding their traditional Pueblo ethos of conformity to the group. Thus, the Picuris may have been predisposed to turn their backs on their pueblo and to take up the life of horse-borne hunters. Whatever the case, Picuris Pueblo never regained its former strength and vigor. From less than 400 following the Ulibarrí mission, the population of the pueblo continued to dwindle to 223 in 1776, and 122 in 1864. Today the resident population of the pueblo is still less than 130.

The decline of Pecos, the other faultline pueblo, took longer to accomplish, but was just as steady. With its population already halved by the stresses of a century of colonization, Pecos consisted of only about a thousand souls when Ulibarrí led his sixty repatriated Picuris through the pueblo in 1706. Pecos then was badly divided. One faction favored the restoration of Spanish authority, probably in hopes that increased civil order would strengthen the pueblo's control over the westward diffusion of trade from the plains. Leaders of this group had saved many Spaniards from slaughter in the 1680 rebellion by warning the governor several days in advance of the uprising. Many other members of the pueblo, however, opposed the Spanish fervently and participated in the bloody siege of Santa Fe. Now that the Spanish were reasserting their colonial authority, the two Pecos factions struggled desperately with each other for control of the pueblo. Ultimately the pro-Spanish faction prevailed, and the victors swiftly murdered or executed most of the leaders of the opposition. Those who escaped followed in the footsteps of the Picuris and went to live with the Apaches. The pueblo they left behind, though weaker than ever, soon emerged as the Spaniards' strongest and most willing military ally.

At the time of the Reconquest, Pecos was able to muster four hundred armed warriors. They were among the the most experienced and proficient fighters in the colony, and Vargas and his successors, holding out the promise of booty, enlisted them to serve as auxiliaries in virtually every military campaign that was mounted in the province through the next fifty years. The constant fighting, of course, steadily depleted the numbers of the troop. Worse, however, was the strain of defending Pecos Pueblo itself.

A new and more deadly enemy, the Comanches, had begun to make their presence felt. Linguistically and culturally related to the Utes, they had migrated from the Great Basin to the northern plains, and thence, pressured by musket-armed Pawnees and attracted by Spanish horses and trade goods, they drifted south to New Mexico, appearing first in Taos about 1705. The Comanches soon upset the balance of power in and around the colony. They drove the Cuartelejo Apaches from their rancherias south to the Texas plains, where they acquired a new identity as the Llanero branch of the Jicarilla tribe. Other Jicarillas, a mountain branch called Olleros, were so harried by the Comanches that by about 1730 they sought the protection of Spanish arms, relocating their rancherias on Pot Creek (Rito de Olla) between Picuris and Taos and in the Rio Grande valley near Velarde. These "tame" Apaches earned their keep by serving the Spanish as scouts and auxiliaries, and they became Hispanicized to the extent that they sometimes appeared as plaintiffs before the Spanish courts.

For the Pecos the advent of the Comanches was doubly unfortunate. First, their Apache trading partners were scattered in every direction. Second, for reasons which today are poorly understood but which may have had to do with the pueblo's long-standing friendship for the Apaches, several large Comanche bands resolved to annihilate Pecos and destroy it utterly. They nearly succeeded.

In the days before the horse came to New Mexico, Pecos had been militarily superior to its challengers, but now the isolated mesilla, with its corn and wheat fields stretching along the Rio Pecos a mile and half away, was glaringly vulnerable to the swooping attacks of Comanche horsemen. Again and again workers stranded in the fields were cut down or kidnapped. The Apache bands who occasionally camped at the pueblo for mutual protection seemed only to elicit the greater wrath of the Comanches. The earthen breastworks that Governor Velez Cachupín ordered built in 1750 and the small garrisons of troops that were sporadically stationed at the pueblo stemmed the Comanche tide only temporarily. The fields by the river were abandoned. Perilously short of food and bereft of trade now that the Comanches controlled the plains, the Pecos teetered on the brink of total ruin. In 1776, as ambitious American colonists sweated under their wigs in Philadelphia, a Mexican cleric, Fray Atanasio Domínguez, counted a mere 269 individuals remaining at Pecos, coaxing the soil around their embattled mesilla to grow rain-watered crops, the fruits of which "do not last even to the beginning of a new year from the previous October, and hence these miserable wretches are tossed about like a ball in the hands of Fortune."

The decline into which Pecos had plunged cannot be attributed entirely to the Comanches. Epidemic diseases scourged the pueblo at fairly regular ten-year intervals, a particularly severe outbreak occurring in 1781 when smallpox claimed the lives of more than a quarter of the population of the colony. Evidently this outbreak of smallpox was the final blow for the neighboring Tano pueblo of Galisteo, which is not recorded as a living community after 1782. Its survivors simply migrated down the Rio Galisteo to Santo Domingo to begin life anew.

Throughout the eighteenth century, people from Pecos had probably been doing the same kind of thing. They were among the most independent-minded people in the Pueblo world. Like the Picuris, they had long been exposed to the individualism of Apache culture; moreover, they possessed a skill that was in great demand all over the colony from Taos to Isleta. They were carpenters. Having been trained in the craft by Franciscans during the seventeenth century, they had been able to keep their skills even through the turbulence of the eighteenth. As late as 1776 Fray Domínguez, who found little else good to say about the pueblo, noted that most of the Pecos were good carpenters, and since at that time skills of any kind were in short supply in New Mexico, proficiency in woodworking implied a high degree of economic mobility. Now that the pueblo was crumbling from without, those who could leave probably did, causing it to crumble still faster from within.

Even after the Comanche threat finally passed, the fortunes

*Photo 11. Wahu Toya, a
Pecos survivor. Photo by
John K. Hillers, 1880.
Courtesy Museum of New
Mexico.*

of the pueblo continued to decline. Trade in plains goods,
instead of being restored to Pecos, followed the now peaceful
Comanches onto the plains in the creaking two-wheeled carretas of Hispanic comancheros. At Pecos in 1815 no more than
forty Indians remained. As a tribe, they faced extinction. Their
particular mosaic of customs and beliefs, which was not rep-
licated anywhere in creation, threatened to disappear. They
clung together as a community long enough to witness the
vigorous expansion of Hispanic New Mexico's population and
territory during the early nineteenth century. And they were
still on watch at their embattled mesilla as canvas-covered
Pittsburgh wagons, driven by American plainsmen, trundled
past Pecos en route to the markets and fandangos of Santa Fe.
But by 1838, they reached the limit of their endurance. The
last seventeen Pecos survivors abandoned their five-hundred-
year-old home to the slow erosion of wind and rain and to
more rapid destruction at the hands of scavenging travelers
on the Santa Fe trail. They bundled up their few belongings
and trekked to Jemez Pueblo on the far side of the Rio Grande,
which was the last place on earth where their language was
spoken.

4

Holding On

*Taking by the hand the said settlers and going
over the tract they plucked grass, cast stones, and
shouted, "Long live the King!" and I placed them
in possession in common, quietly and peaceably
without prejudice to third parties.*
> —Juan José Lovato, Alcalde and War
> Captain, on the founding of Las
> Truchas, April 24, 1754

THE EIGHTEENTH-CENTURY Spanish colony in New Mexico was
very different from the seventeenth-century colony. Whereas
formerly the Spanish had built their society on the backs of
the Pueblos, in the 1700s they built it beside and among the
natives. They had learned from the Revolt of 1680 that they
could not forcibly impose their religion on the Pueblos, nor
exploit the Indians' economic energies as relentlessly as they
had done before. And so what emerged in the eighteenth century was not a European society supported by an agricultural
and laboring class of Pueblo Indians, but a society that had
changed its silken clothes for coarse woolens and had found
its new foundation within itself—in a broad class of Spanish-
speaking, though hardly Castilian, peasants.

The cultural diversity of contemporary New Mexico owes
much to the change in structure of colonial society. Now the
Pueblos and the colonists were allies in an economic sense,
working together rather than at each other's expense, even as
the depredations of Comanches and other hostile tribes required them to ally themselves militarily. Certainly the presence of mutual enemies was a major factor in bringing about

the change, but it was not the only one. Probably an even more important reason for the diminution of fighting within the colony was that the Spanish authorities had finally come to understand that New Mexico offered little worth fighting for. The colony was geographically isolated; its mountains had yielded little worth the effort of mining; and in general it was too rugged, too impoverished, and too deficient in marketable resources to warrant any kind of aggressive development.

Stretching narrowly along more than a hundred miles of the Rio Grande valley, the Spanish colony of New Mexico did not have to thrive in order to fulfill its royal purpose; it had only to exist—yet even that seemed at times an impossible task. During the eighteenth century, as depopulated Picuris struggled to survive and Pecos entered its final decline, the ancient dream of a golden Quivira, of a *new* Mexico to match the city of the Aztecs, appeared ludicrous. New Mexico was no more than a minor military prize, which by belonging to Spain served the exalted purpose of not belonging to any of Spain's enemies.

Economically the colony was of negligible value to its mother country. Each year it sent south a few thousand bushels of piñon nuts, a train of creaking carts loaded with raw hides and buffalo robes, some salt, coarse blankets, and an occasional surplus of Indian slaves which the colony could not absorb. Usually these unfortunates were captives obtained from other Indians, and the majority of them were destined to live out their suddenly shortened lives in the mines of Chihuahua and Parral. Whenever the colony had respite enough from Indian raids to develop its herds, it also exported its long-legged goat-like churro sheep, but in the second half of the eighteenth century, those periods of tranquility were few and far between.

The colony was pressed hard from all sides by Comanches, Navajos, Utes, and Apaches (excepting the Jicarilla). Because it produced little that was worth the long haul from Santa Fe to Chihuahua, it was seldom able to import much in the way of essential commodities like firearms and metal tools, let alone luxuries like sugar, coffee, and fine clothes. With a struggling population at midcentury of between 7,000 and 8,000 Hispanos and somewhat fewer than 10,000 Pueblo Indians,

New Mexico was a tired backwater. The viceregal authorities in Mexico, who expected their dominions to pay their own way, were seldom inclined to send soldiers, guns, horses, or anything else of value through the deserts to Santa Fe.

The king in Madrid and his xenophobic administrators in Mexico had use for New Mexico only as a bulwark against the French, whose presence on the plains the Cuartelejo had reported to Ulibarrí in 1706. Until 1766 when Great Britain defeated France in the French and Indian War, the Spanish expected the French to attempt westward expansion from the Mississippi and to hunger for possession of the rich silver mines of northern Mexico. The other principal justification for maintaining the colony was, as ever, that the Church could not turn its back on the Christianized Pueblos, nor even on the thousands of pagans who had yet to be converted. And so the colony, which would have withered completely if left to its own devices, received enough royal nourishment to stay alive. Like a beleaguered desert plant, it had one long thin taproot that reached to Mexico and a scant few leaves in Albuquerque, Santa Fe, Santa Cruz, and Taos.

One may gain some idea of the isolation from which New Mexico suffered in the eighteenth century from the fact that Fray Atanasio Domínguez, a Franciscan cleric who was charged with the important task of making an economic and ecclesiastic survey of the province, was obliged to wait for *five months* at the site of present-day El Paso–Juárez before a sufficiently large and well-armed wagon train could be assembled to attempt the trip north. While Domínguez cooled his heels through the winter of 1775–76, he had plenty of time to reflect on the raids of hostile Apaches along the *camino real* that had almost entirely severed communication between Santa Fe and Mexico. Although he probably did not understand that the Comanches had caused the Apaches' desperate acts by driving them from the southern plains, the Comanches were well known to him by reputation. Their attacks on the settlements of New Mexico threatened the very life of the colony. In a letter to his immediate superior, Fray Isidro Murillo, he stated with his usual frankness that the raids of the nomads had

Photo 12. Comanche men, probably about 1880. Courtesy Museum of New Mexico.

reduced the population of New Mexico to "a state of miserable panic."

Comanches were the preoccupation of all of New Mexico. For three and a half decades after arriving on the southern plains, they kept a more or less steady peace with the Spanish colony and its Pueblo allies. But about 1740, the balance of regional power began to change. At that time Frenchmen striking west from the Mississippi reached the plains in numbers large enough to wean the Comanches from their economic dependence on New Mexican trade fairs. The Frenchmen offered good guns for Spanish horses, and the Comanches, realizing that with enough guns, horses would always be easy to steal, turned upon the colony with a ferocity that they had until then reserved for the Apaches. Even after the French turned over Louisiana to Spain in 1763 and abandoned the plains, the situation did not improve, for British and American traders soon replaced them as purveyors of arms to the Indians.

By 1776 the flow of arms across the plains was so substantial that Comanches occasionally sold their surplus guns to the miserably equipped New Mexico settlers.

New Mexico slowly shriveled as one decade of Comanche warfare gave way to another. The colonial authorities never assessed the exact extent of the colony's retreat, but it is known with certainty that the unhappy settlers of Abiquiu and Ojo Caliente, who defended the vulnerable northwest frontier of the colony, abandoned their plazas in 1747; that in spite of being resettled three years later, the Ojo Caliente valley was again deserted in 1769; and that the valley was still deserted or had been deserted again when Don Juan Bautista de Anza, the brilliant tactician who finally delivered New Mexico from its Comanche oppressors, passed through it in 1779. The main Spanish settlement of Taos, meanwhile, was deserted in 1770 when its inhabitants moved into the nearby pueblo for protection, and in 1782 the village of Embudo, which lay well within the frontiers of the colony, El Nacimiento (today's Cuba),

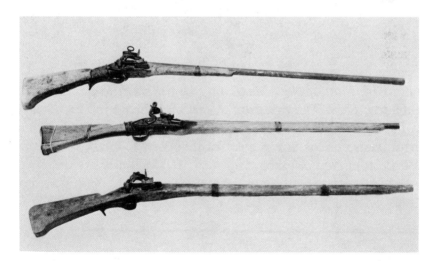

Photo 13. Spanish colonial firearms. Courtesy Museum of New Mexico.

Photo 14. Map of New Mexico by Don Bernardo de Miera y Pacheco, 1779. The notation describing the abandoned valley of the Rio de Ojo Caliente reads "Poblaciones arruinadas por los Enemigos Cumanches." *Courtesy Museum of New Mexico.*

and some ten other settlements were similarly abandoned. The colony was crumbling at its edges, and refugees jammed the main towns.

Paradoxically, this dark period produced an actual expansion of Hispanic settlement in the southern Sangres. In 1747 young, brash Tomás Vélez Cachupín became governor of the colony and began to accomplish what none of his predecessors had been able to do: to achieve peace with New Mexico's nomadic neighbors by orchestrating a complex system of truces between the various warring groups. In addition to his treaties, which unfortunately did not outlive his own tenure in office, Vélez Cachupín endeavored to strengthen the frontiers of the colony by establishing fortified settlements athwart the main routes that the Comanches used for their raids. Two such settlements were the mountain villages of Las Trampas (1751) and Las Truchas (1754), whose purpose was to protect the populous Santa Cruz Valley from invasion through the high country of the Sangres. The settlers who received the royal land grants that Vélez issued for these places were a hardy and, some would say, suicidal lot. They were obliged to wrest their homes from a cold mountain wilderness and to defend themselves often with only lances, bows, and arrows against better-armed Comanches. Their villages, to use a term from jungle wars in Burma and Vietnam, functioned as "strategic hamlets"; they were forward redoubts, a challenge to the enemy. Unlike some of their recent counterparts in Southeast Asia, however, Las Truchas and Las Trampas served their purpose well. Vélez ordered that each of them be built as a fortified compound with an open space within where animals could be driven during times of danger. In the case of Las Truchas he specified that the village be "a plaza in a square with only one entrance for carts so that the inhabitants may defend themselves against the attacks and assaults of their savage enemies." Helped by their defensive design, the villages survived the Comanche wars, and in some ways, even prospered.

It has been said that to be Hispanic in northern New Mexico is to be "of a village," and Las Truchas and Las Trampas are the quintessential Hispanic villages of the region. Born in adversity and isolated by the ruggedness of their environment, they were places where sufficiency gave way easily to suffering and where arduousness and danger were the common conditions of life. The people who settled them were a resilient breed. Passionately Spanish and Catholic in their cultural orientation, they were descendants as much of the New World as of the Old.

The *pobladores* of Las Trampas consisted of a dozen families, almost all of whom were related in some way to the group's sixty-year-old leader Juan de Arguello. They were recipients of the Crown's generosity not because of any special merit but because, as Vélez Capuchín noted in the instrument of their grant, "there is not [in Santa Fe] land or waters sufficient for their support, neither have they any other occupation, trades, or means of traffic, excepting agriculture and the raising of stock." They were simple, unlettered people, a mixture principally of Spanish and Mexican Indian stock, and like their former neighbors in the lower-class Barrio de Analco on the south side of the Santa Fe River, they had always been poor.

By 1776 the population of Las Trampas had increased more than five fold, to sixty-three families with 278 persons. The fastidious Domínguez described them as

a ragged lot, but there are three or four who have enough to get along after a fashion. They are as festive as they are poor, and very merry. Accordingly, most of them are low class, and there are very few of good, or even moderately good blood. Almost all are their own masters and servants, and in general they speak the [bad] Spanish I have described in other cases.

"Blood," as Domínguez called it, featured remarkable variety in eighteenth-century New Mexico. No fewer than seven racial categories were used to classify the population. At the top of the social hierarchy stood the *españoles*, nominally the true Spaniards, who among themselves distinguished between *peninsulares* or *gachupines*, born in Spain, and *criollos* (creoles), born in the colonies. As in every stratified society, membership in the upper classes often depended less on ancestry than on affluence, authority, and friendship. This was especially true of the next class, called *color quebrado*, literally "broken color," a term applied to citizens of high standing who were yet too dark or Indian-featured to be accepted by the *españoles* as equals.

In Albuquerque and most other New Mexican towns and villages, with the notable exception of high-class, presidial Santa Fe, *mestizos* made up the majority of the population. These people, a mixture of Spanish and Mexican Indian descent, were considered distinct from *mulatos*, of mixed Spanish and African parentage, and from *coyotes*, who were the offspring of Spaniards and New Mexican Indians. (Interestingly, the term *coyote* nowadays refers to people who are part Hispano and part Anglo.)

Two more classifications were used to distinguish the full-blooded Indians of the colony: *indio* referred to those who retained their tribal identity, while *genízaro* (collateral to the English word *janissary*) described a large and ever-growing class of mainly non-Pueblo Indians who had either abandoned or been stripped of their tribal associations and who had adopted an ostensibly Spanish way of life. Such villages as Abiquiu, Ojo Caliente, and later Anton Chico were settled almost exclusively by *genízaros*, probably a majority of whom were survivors of the colony's brisk and carnal trade in slaves. At all the trade fairs at Taos, Pecos, and Santa Cruz captive women and children, selling as slaves for a couple of horses a piece, were one of the most sought-after commodities. Paiutes brought in by Utes, Pawnees brought in by Comanches, Apaches brought in by nearly anybody all found a ready market. It was a poor man indeed, as poor as a Trampero, who could not afford to buy one of these unfortunates to assist his wife with household chores. The official justification of this traffic in human beings was that it brought new souls into contact with the Word of God, but embarrassment could result, and sometimes it did, when a ten-year-old boy, duly purchased from a Comanche, turned out to be a baptized Spanish-speaking citizen of Sinaloa. Then the buyer was obliged to balance economics against humanitarianism, which not all buyers found easy to do. Slavery in New Mexico, however, was an infinitely gentler institution than the slavery practiced in the young republic of the United States, far to the east. Children of slaves were always considered to be born free, and the slaves themselves, usually obtained at an early age, were generally given their freedom soon after reaching adulthood.

Although cumbersome, the class system of the colony was far from rigid. An analysis of Albuquerque data from the census of 1790 reveals that fully 69 percent of the marriages in the *villa* were between members of different racial classes. At so rapid a rate of miscegenation, the elaborate system of categories could not and did not hold up. Before long, thanks also to the rise of democratic nationalism in Mexico, everyone who spoke Spanish as a first language and was not an *indio* was considered simply to be *mejicano*.

Out of the ethnic melange of New Mexico came the mestizo settlers of the land grant named Nuestra Señora del Rosario San Fernando y Santiago, which lay in the mountains a few miles south and east of the Las Trampas grant. They called their village Las Truchas, and their experience defending it from the Comanches was typical of the entire colony's struggle. Mostly Espinosas from Chimayo and Romeros from Santa Cruz, they built their village on a flat, dry shoulder of the mountain range. With great effort they brought an irrigation ditch from the base of the Truchas Peaks, around the side of an intervening mountain, and then down a long canyon to the village. Considering that they had no metal shovels or picks with which to dig the ditch, much less a transit to lay it out, their achievement in bringing water through a half-dozen miles of precipitous mountain country to their fields and homes was prodigious. With their *acequia madre* in place, they could at last settle down to the more ordinary business of building homes and growing wheat (maize was risky at that altitude). Unfortunately, no amount of engineering or hard work could eliminate their exposure to the Comanches, and they had not been long installed in their new village when the war with the nomads took a decided turn for the worse.

For a long time the Comanches had treated New Mexico with a certain degree of care. They pillaged in one place and traded their booty back in another, never ruining their good fortune by destroying too much of the colony's productive capacity. In the 1770s, however, a chief named Cuerno Verde, whose name derived from the green-painted buffalo horn he wore in the center of his headdress, ascended to the leadership of a major portion of the Comanche nation and dedicated himself and his people to the utter obliteration of the colony. Raids grew more frequent, casualties increased, and the people of Truchas, isolated on their lonely frontier, felt the pressure as much as anyone. In 1772 they wrote to the governor, Pedro Fermín Mendinueta, alerting him that the mountains behind

Photo 15. A view of Truchas and its log barns by John Collier, January 1943. The Jemez Range, Chicoma Peak at center, stands in the distance. Courtesy Library of Congress.

the village were full of Comanches and that they expected imminent attack. Desperate, they pleaded for help:

We beg your Lordship to give us twelve firearms with powder and shot with which we may defend ourselves and our families, and we beseech your Lordship immediately to instruct the Alcalde Mayor and War Captain of this district not to conscript us for campaigns, invasions, or rescues so as not to leave this place abandoned and ripe for ruin. And we also ask that Indians from the Pueblos be given us as lookouts and scouts of the mountain passes so that we may have the earliest possible warning in order to prepare for attack.

Governor Mendinueta, who, although not as creative or skillful as Vélez Cachupín, was nonetheless an intelligent and forthright leader, had no choice but to deny all the requests. At that time there were only about six hundred muskets and a hundred and fifty pairs of pistols in the entire colony. The shortage of firearms was so severe that a strange reversal of

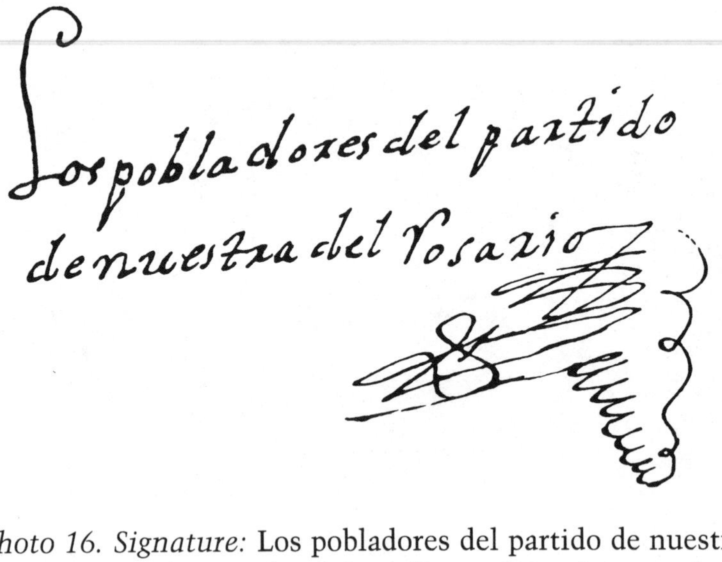

Photo 16. Signature: Los pobladores del partido de nuestra del Rosario—*the people of the village of Truchas, on the Rosario land grant. Courtesy State Records Center and Archives.*

Photo 17. Signature of Pedro Fermín de Mendinueta, governor of New Mexico. Courtesy State Records Center and Archives.

roles between colonists and Indians had developed: at the trade fairs Hispanic settlers had been obtaining guns from the Comanches since as early as 1760. Mendinueta and his predecessors tried to remedy the situation by appealing to Mexico City for guns, powder, horses, and soldiers, but especially since 1762, when France ceded the Louisiana Territory to Spain, their appeals had gone largely unanswered. If there was no European threat to New Mexico, then the colony had no strategic importance in the grand scheme of empire. It would have to fend for itself.

As he had no guns to give the Trucheros, Mendinueta had no Indians to give them either, nor could he exempt the settlers from their obligation to take part in government campaigns. He pointed out in his letter to the settlers that the Pueblos, like everyone else, had crops to tend and families to defend and that every frontier settlement in the kingdom wanted to be spared from contributing men to the militia. He told them that rather than plead for special treatment, they should guard their own roads and trails and bend every effort "to finish the plaza and torreones with which you will be secure even if attacked."

Apparently the Trucheros had either lagged in building their fortified village, or else they had been prevented from com-

[73]

Photo 18. *Signature of Cristóbal Montes Vigil, Alcalde and War Captain of Santa Cruz and Marselino Vigil and Rafael de Medina, witnesses. Courtesy of State Records Center and Archives.*

pleting it by raids, conscriptions, or some other kind of difficulty. The torreones Mendinueta mentioned were the main battlements of the tiny fortress, being thick-walled towers of logs and adobe in which the village's defenders hoped to be invulnerable. Soon enough Mendinueta would have cause to remember the Trucheros and their torreones, but for the moment, he instructed the alcalde of the Santa Cruz district to inform them of his decision. The alcalde, in his turn, wrote his response to the order on the bottom of the same sheet Mendinueta had used (there being at the time a severe shortage of paper in the colony). Thrice he repeated, each time in increasingly tortured and circular language (a style not uncommon among officials of our own day) that he would tell the settlers what the governor said. As simply as any other relics from the period, these three messages bespeak the agony of New Mexico during the darkest hours of Comanche terror. The letter of the settlers was riven with fear, that of the governor austere and unhopeful, and the response of the alcalde almost wholly irrelevant. The government, it seemed, could preserve protocol but not its citizens.

Reflecting on the tribulations of this period, historian Alfred Barnaby Thomas has written that it would seem difficult to find elsewhere in the west an equal record of suffering on the part of those pioneers from Indian attack. The fact that the New Mexicans maintained their hold in the face of critical neglect, developed as far as they could their own defense, and thus resisted successfully the effects of Indian invasion is testimony to the hardy and vigorous character of the New Mexican people.

Through their folklore the people of New Mexico have said roughly the same thing. The Comanche wars spawned scores of ballads and legends as well as a major folk-drama called, not surprisingly, *Los Comanches*. The folk materials give some idea of what the men and women who endured those miserable decades felt about them and they also shed light on the significance that succeeding generations found in the experience of their forebears. Especially in the latter respect, a tale that was born in Truchas in 1774 and still lives in oral tradition seems a useful mirror for the past and a fitting close to this account of colonial perils.

The events that gave rise to the tale began in the fall of 1774 when Governor Mendinueta made plans for a punitive expedition against the Comanches, who that year had inflicted heavy losses on Pecos, Picuris, Nambe, and other settlements. Unlike some of his less competent predecessors, Mendinueta took steps to ensure that the Comanches should not learn ahead of time of his intentions. He ordered the alcalde of Taos, where the Indians came regularly to trade and, not incidentally to gather intelligence, to seize any Comanches who appeared there.

On September 19, eight Comanche men and eleven women entered the town and were duly captured. The next day they were marched to Las Truchas, halfway along the mountain trail to Santa Fe, and were locked up in a torreon. The night was bitterly cold. As Mendinueta later reported to Viceroy Bucareli in Mexico City,

In the evening two careless and stupidly humane guards entered the tower to build a fire for the Indians, who killed them both with their own weapons and took possession of the entrance and the rest of the arms nearby and with them defended themselves all night.

They wounded many settlers. Having been asked repeatedly to surrender, they scornfully rejected the demands, saying they wished to die fighting. The besiegers, fearing that they would escape or kill others, set fire at daybreak to the tower. In view of this, the Comanches still rejected offers to give themselves up as prisoners. They elected, before being barbarously choked to death with smoke, to sally forth with arms in their hands to defend themselves to the last. Accordingly, they did not succeed in giving any news to their tribe.

Over the next 198 years, while Mendinueta's letter grew yellow and brittle in archival files, the descendants of the people who fought that battle kept alive a memory of it. Slowly, as the event grew more distant in time, the story of the fight became less a device for preserving the past than for entertainment in the present. Some particulars were forgotten, others embellished according to the teller's needs and art, but the pivotal action remained the same: Indians were smoked out of a building in Truchas that had become both their fortress and their jail. It is not a story that everyone in Truchas knows or even has heard, but white-haired Juan López knew it, and on a relaxed Sunday afternoon in 1972, with several guests gathered in his living room, he saw fit to tell it.

He began by explaining that long ago all the houses of Truchas were connected to make one large building, which could be closed up with all the villagers and their animals safe inside it in the event of Indian attack. Those were hard days, he said. You had to be constantly on guard against the Indians. They were always jealous of what the Spanish people had in their villages: good houses, horses, plenty of meat and grain, sometimes some wine.

One day the Indians caught the people of Truchas by surprise. They said, "Now you have to leave. We are taking your village. If you don't go away and let us have this place, we will kill all of you."

Well, said Juan López, it was very clear that the people of Truchas had to do what the Indians told them. They went away, but they did not go far. What the Indians did not know was that the Trucheros were prepared for just such an event

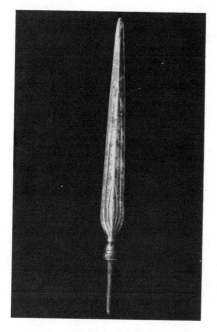

Photo 19. Spanish lance point. Photo by Arthur Taylor. Courtesy of Museum of New Mexico.

to occur. When they had built the village years before, the men of Truchas had been smart enough to put strong locks on all the doors and windows. The peculiar thing about those locks was that they could be securely fastened not just from the inside, but from the outside of the building as well.

That night while the Indians feasted and drank and celebrated their victory, the people of Truchas crept back through the darkness to their village. Silently they threw the locks and barred the windows and doors so that the Indians could not get out. Then a few of the men climbed onto the flat roof of the building, and *ristras* of chile were handed up to them.

It was a cold night and the Indians had made a fire in every fireplace. The men on the roof separated and each one very quietly tiptoed over to a different chimney and dropped his chiles down. Then he put a board over it to keep the smoke from coming out, and waited.

The effect was like tear gas. As anyone knows who has

roasted chiles, the pepper releases a terrible gas when it burns. It irritates the eyes, nose, and mouth; it burns the throat and lungs. Pretty soon the Indians were in a panic. They ran for the doors, but they could not open them. They ran to the windows, but there was no way to get air. The smoke of burning chiles filled the rooms, and the Indians were blinded, choking, unable to breathe. They screamed to be let out.

The Trucheros said, "Will you do what we say?"

The Indians had to say yes.

The men opened a place for the Indians to throw out their weapons, and once that was done, they told the Indians that they would let them live only on the condition that they go away and never make trouble for the village again. The Indians, who were still coughing and choking from the smoke of the chiles, were only too happy to agree, and they swiftly disappeared into the night, never again to be seen or heard of in those parts.

"And that is how the people of Truchas outsmarted the Indians and won back their village," Juan López concluded. "Maybe it is not all true, but that is the way I heard it."

As the story suggests, there was no outside agency to which the Truchas settlers might have appealed for help. In order to survive they had to rely solely on their own resourcefulness, their cunning, and—this may be the meaning of the story— on their special character as a people. Chile is the emblem of the Hispanic *poblador*; it is the trademark of his culture. In Juan López's story it comes to the rescue of Truchas, releases its nearly magical properties for the benefit of the people, and restores them to their rightful home. Such a story is the raw material of myth. It expresses the independence, resilience, and pride that time has forged into the heart of New Mexican culture.

5

Lithic Parallels

*On the side of Jicarita Peak there are caverns
similar to the Carlsbad Caverns, but these
caverns have not been fully explored. It is only
known for sure that some shepherds have
penetrated into them for about a mile, and they
say they found stalactites within.*
       —Simeon Tejada, NM WPA Writers'
       Project, 1939

ULTIMATELY, THANKS TO THE LEADERSHIP of Don Juan Bautista
de Anza, the New Mexicans dealt a crushing blow to the Comanches in 1779 by soundly defeating their leading chief Cuerno
Verde in battle. The years following that victory, however,
were not a period of peace. Not all the Comanches agreed to
the treaties that the Spanish arranged with the tribe, and those
who did upheld them fitfully. Meanwhile the Apaches (ex-
cluding the Jicarillas) continued to beset the colony with all
their energy, as did sporadically the Navajos and Utes. In truth
all of New Mexico was still a frontier where the perils of
wilderness life reached to the front door of every house.

Nevertheless, conditions had improved. Men traveled more
freely. Herders could take their animals farther afield. Bands
of ciboleros could ride out past the protective barrier of the
mountains, and using lances as the Comanches did, hunt buffalo on the eastern plains. The danger of attack remained great,
but it no longer outweighed the value of a winter's supply of
meat and a cartload of robes.

The colony had grown in spite of its hardships, and new
settlements sprang up to satisfy the pressure of increased pop-

ulation. Some, like Ojo Sarco (clear spring) between Truchas and Trampas, were established within existing land grants, while others, authorized by charters of their own, opened new lands. One new land grant settlement was made in the name of Santa Barbara, who, appropriately enough, was considered to be the patroness of those who faced danger and sudden death. The Santa Barbara Grant came into being on April 3, 1796, when Valentín Martín and forty-one fellow settlers took possession of the broad mountain valley, ideal for growing wheat, through which the Rio Santa Barbara leaves the present Pecos Wilderness. The settlers must have prayed that Santa Barbara stand by them, for their land, under a previous grant, had been abandoned by its owners, probably at the same time in the early 1770s that Comanche raids a few miles down the valley forced the Picuris to tear down their mission church and move it, brick by adobe brick, to a more defensible location.

The new communities made settlement denser; travel between communities became surer and easier, and the fragile economic bonds that held the colony together were strengthened. New Mexico's sheep herds rapidly expanded and may have grown to include more than two million animals by the mid-1820s. More important, however, were the new settlements that rolled back the edges of the Spanish world. In 1794 a band of a little more than fifty men, who boasted possession of twenty-five firearms among them, pressed east from Pecos to establish the village of San Miguel del Vado at the ford where the main trail to the plains crossed the Rio Pecos. At more or less the same time but far to the north, pioneers from Trampas, Embudo, and Picuris audaciously breached the mountain frontier by hauling their belongings up the Rio Pueblo, through the same pass that the Comanches had used countless times for invasion, and descended the steep eastern side of the range to settle in a place they called Lo de Mora.

It was probably also at this time, as their villages proceeded slowly to encircle the southern tip of the Rockies, that Spanish-speaking people first began to travel and camp regularly in the high country of the Sangres. To be sure, Hispanic prospectors, beginning as early as Oñate's time, had repeatedly

Photo 20. Artist's view of Mora, New Mexico, c. 1858. East Divide of the Sangre de Cristo Range in background. Courtesy Museum of New Mexico.

scouted the mountains for signs of mineral wealth. Even in the 1760s, during the height of the Comanche wars, Governor Vélez Cachupín had seen fit to send a prospecting expedition to the San Juan Mountains in what is now southwest Colorado, and so we can safely assume he had already satisfied himself as to the nearby Sangres' potential, or lack of it. (Interestingly, Vélez's expedition returned with some black, ore-bearing rocks, but no one in Santa Fe was sufficiently knowledgeable to determine what kind of ore the rocks contained.)

But other, poorer prospectors surely did not give up on the Sangres and must have worked their way little by little up the cold mountain streams panning the sands for yellow glitter. Hunters, too, would have climbed the slopes, for not even the rich hauls of the ciboleros could have entirely satisfied New Mexico's growing population, and in those days the meat of the mountain sheep was considered the greatest culinary delicacy in the region. The occasional vaquero, chasing a lost cow or horse, might also have pushed into the mountains far enough to see the land merge with the sky at the top of some high divide, and like the prospector or hunter before him, he might have gone the little distance that remained, just to see what was there.

What was there is still there, and we can see it today almost as those Hispanic pioneers saw it. From any high vantage point we would see a dark and choppy sea of forests from which immense alpine islands emerge, windswept and treeless and so blasted by sun and wind that the richest color they wear is a tawny, flesh-hued brown. The bald islands twine through the distance, alternately swelling into lofty peaks and subsiding into low ridge saddles. They twist together making a kind of wild yet intricate script on the land, a script that tells the history of the land, if one knows how to read it.

The huge natural amphitheaters, hollowed into the sides of the alpine masses, were the seats of the great Pleistocene glaciers that carved the mountains to their present form. On the floors of these amphitheaters, which the pioneers called rincones and geologists call cirques, grow the deepest grasses and the dankest forests in the range, and in some of them lie the small, clear lakes, suffused with the turquoise of the sky, that the Pueblos revered as the homes of spirits. These lakes are all that remain of the giant depressions that the weight of the glaciers carved from the earth, and in many cirques where the old lakes have filled in, their outlines may still be read in the darker grasses of the bogs that have replaced them.

The glaciers left a varied legacy in the high country of what is today the Pecos Wilderness. From anywhere above the timberline where the view takes in the whole of the range, it is easy to see that the effect of the ice on some mountains was different from its effect on others. All along the western edge of the high country the peaks are craggy and austere; they rise into sharply pointed summits, aglint with the bare rock of cliffs and narrow, razor ridges. But away to the east the peaks have smoothly rounded shapes. They appear soft where the mountains to the west are hard, female where the others are male. And the contrast between the landforms is not confined to their skylines. In every respect, the land below the eastern mountains is more fecund and less severe than the land in the west. The two halves of the high country contrast to each other as yin to yang, and the essential difference between them, which underlies the whole character of the land, resides in the materials that compose them.

The oldest materials in the range, the granite, quartzite, amphibolite, and schist that compose all the high peaks along the western divides, were formed in Precambrian times, about 1.6 billion years ago. Their most salient common characteristic is their hardness, and it affects every aspect of the land that they form. Hard rock means slow soil formation and steep topography. The steepness is critical; it works against the formation of meadows and the retention of water and soil. It leads in the end to a landscape that is dramatic and challenging but biologically stingy.

The entire western third of the Pecos Wilderness is built out of Precambrian rock, mostly granite and quartzite. Its thin soils and low production of grasses and forbs make it poor for deer and elk compared with the rest of the high country, but by its ruggedness it epitomizes the longevity and endurance of the bedrock from which it is made. Remarkably, for more than a billion years, the Truchas Peaks (or more accurately, the landforms from which they are descended) have continuously been among the highest points of land in the region. Not once but many times in their long past they have been islands barely topping the sea or low hills above a plain. And again and again they have materialized as mountains and endured new cycles of downwearing and resurgence.

For more than a billion years after the formation of the Precambrian quartzites and granites of the Sangres, no new rocks were laid down, or if some were, then subsequent erosion removed them entirely. The resultant gap in the geologic record ends with the formation of a thin stratum of shales, which are said to rest "unconformably" on the Precambrian basement rocks.

These rocks formed during Mississippian time, which marks the first half of the so-called Carboniferous Age, when mankind's family fortune of oil, gas, and coal accumulated under shallow seas. It was in the second half of the Carboniferous Age, during Pennsylvanian time, that a vast mountain highland, called the Uncompahgre Range, rose up over much of north central New Mexico, stretching far into Colorado. Its southeastern limit was a great wall, more than a thousand feet high, standing where the western crest of the Sangres stands today.

For fifty million years the lowlands east of the wall were either choked with swamps or flooded by shallow seas, and whole mountains of erosional debris washed down from the highlands to mix with the water-logged sediments below. Layer upon layer was trapped and crushed under the weight of suc-

Photo 21. The Pecos High Country looking west from atop East Divide. Pecos Baldy on left, Truchas Peaks center. Author photo.

ceeding deposits and advancing seas. The landscape had become a kind of rock press, a stone factory. From the mud of bogs and lake bottoms, shale was formed. Dunes and river sands produced sandstone, and the bones and shells of marine animals were transformed along with other materials into limestone.

In some parts of the Pecos country the deposits of the Pennsylvanian period reach thicknesses greater than a thousand feet. They cover the old Precambrian bedrock (with the thin stratum of Mississippian rock sandwiched between them) like a great tan blanket, and only where the blanket is slightly torn—along some of the larger rivers, where torrents of water cut box canyons—do the older and harder rocks show through. Some of the tallest and most massive of the Wilderness peaks are built from this rock. Round-domed Jicarita, a mountain sacred to Picuris, is one of them; Santa Barbara Baldy is another.

Santa Barbara is the hub of the high country wheel. It is against her sides that creeks belonging to all three of the mountains' main watersheds take form: the Rio Santa Barbara, which flows to the Rio Grande, the Rio la Casa, which wanders ultimately to the Mississippi, and the Rio Pecos itself. The Pennsylvania rocks of Santa Barbara make her a fat frump of a mountain. The valleys below inherit the mountain's rounded shapes and her limestone and sandstone soils, which, although not as fertile as the volcanic soils of the Jemez Mountains, are still rich compared to the thin, grainy stuff of the Precambrian country. The slopes around Santa Barbara are steep, but not so steep that they prevent animals from moving over them freely. In the fertile, well-watered grasslands around the base of Santa Barbara, individual herds of as many as two hundred elk congregate every year during fall rut. A few decades ago, however, those same grasslands held concentrations of many thousands of other animals. Through the last half of the nineteenth century and until relatively recent times the lands around Santa Barbara were chronically overgrazed by domesic livestock. The years of misuse are commemorated in the name of one of the glacial cirques cupped into the mountain's side. The cirque is called the Rincon Bonito (pretty corner) on ac-

Photo 22. The Rincon Bonito, looking east to the Great Plains. Twelve thousand years ago a glacier occupied this natural amphitheater. Author photo.

count of the profusion of wildflowers that grow there. And the reason so many wildflowers grow in the Rincon Bonito is that tens of thousands of sheep, year after year, so efficiently removed the grasses from the cirque that inedible wildflowers were all that were left to cover the ravaged land. Bonito, indeed.

If you can visualize the geologic strata of the southern Sangres, you can also visualize the stratifications of the cultural history of the region. Accept for a moment that the cumulative centuries of Desert culture, Anasazi, and Pueblo prehistory form the "basement rocks" of the human record. On top of these, but separated by a great unconformity, lie the experiences of Spain's seventeenth-century colony. They make a stratum that is radically different from the basement rocks in every respect, but it is a thin stratum, as thin as the Mississippian shale of the Sangres, for the Pueblo Revolt had the effect of eroding away much of Spain's early imprint on New Mexico.

The heaviest and most pervasive impacts of Hispanic culture on New Mexico accumulated through the eighteenth and early nineteenth century and forever changed the cultural topography of the region. As great as they were, however, they did not bury all of what had gone before. The hard, resistant Indian "peaks" of Taos, San Juan, and other pueblos still rise today above the rest of the cultural landscape.

The major contours of the land had not yet been shaped, however. It remained for trappers and traders from the east to come as the vanguard of a cultural advance that moved into New Mexico with tectonic force. Soon a new period of transformation began, and if there is a geologic analogy for it, it is the so-called Laramide Revolution, which produced the uplift of the Rocky Mountains, including the Sangres, beginning about seventy million years ago. As the Laramide Revolution folded and crumpled and buckled the Sangres, making them a mountain range, so the Anglo Revolution in New Mexico exerted tremendous pressures on Hispanic and Pueblo societies and eventually produced the broad outlines of the New Mexican world we know today.

6

Pandora's Scouts

"Oh we are approaching the suburbs!" thought I,
on perceiving the cornfields, and what I supposed
to be brick-kilns scattered in every direction.
These and other observations of the same nature
becoming audible, a friend at my elbow said, "It
is true those are heaps of unburnt bricks,
nevertheless they are houses—*this is the city of*
Santa Fe."

—*Josiah Gregg, 1831*

THE BUCKSKIN-CLAD TRAPPERS who lounged in the plazas of Santa Fe and Taos in the 1820s were the first swell in a sea of change that would rock New Mexico for the next hundred years. They were of both frontier French and Anglo descent, and the papers they presented to the Mexican customs officers explained that they hailed from places called Tannesi, Ilanoy, and Misuri. Almost all of them claimed to be citizens of the United States of America, which, if the trappers were a fair sampling, was a nation of adventurous, dirty infidels.

They had come to New Mexico in order to trap beaver, an animal that held no interest for the New Mexicans. Coarse furs had always been a staple on which the colony depended for clothing, bedding, rugs, and other furnishings, and one could make a good profit shipping the hides of deer, elk, and bison down the Chihuahua Trail to the mining provinces of northern Mexico. But fine furs from beaver and river otter were useless. No one in New Mexico knew how to work them into shoes or hats, and there was no market for them in the warm cities of Mexico.

During the last decades of Spanish rule it was well known

Photo 23. An adobe dwelling, Pecos, New Mexico. One half of a stereo view by Ben Wittick, c. 1880. Crude houses like this one shocked early Anglo visitors to New Mexico. Courtesy School of American Research Collections in the Museum of New Mexico.

that foreigners, especially Frenchmen and Anglos from the United States, paid handsomely for beaver pelts, but the xenophobes who ruled in Madrid and Mexico City viewed foreign trade as anathema. Commerce across the borders of the Empire, they felt, led to undesirable competition with Spanish and Mexican trade goods. Worse, it permitted rival powers to learn firsthand just how undermanned and vulnerable frontier provinces like New Mexico really were. It was also widely understood that the isolation of the people from new ideas helped to keep them, if not politically content, then at least quiet.

For these reasons the New Mexico authorities held out no welcome to the Mallet brothers, Pierre and Paul, who made their way to Taos from across the Plains in 1739, or to the traders who followed them in the course of the next several decades. Most of the intruders were arrested and their goods confiscated, but their treatment was not necessarily harsh—at least at first. Two members of the Mallet party liked New Mexico well enough to stay behind and settle there. One of them, however, whose name in Hispanicized form was given as Luis María Mora, became known as a sorcerer and was eventually sentenced to be executed in the Plaza of Santa Fe.

Border tensions eased in 1762 when Spain acquired Louisiana from France, but then increased in 1803 when the Territory became property of the United States. In 1807 Lieutenant Zebulon Pike of the U.S. Army wandered into New Mexico at the head of a small band of explorers and claimed to be lost. He was promptly arrested and accused, probably rightly, of being a spy, and after dining with nervous New Mexican officials in Santa Fe, was hustled off for interrogation in Chihuahua (which enabled him to gather more information than he could ever have reasonably expected).

In spite of the authorities' most conscientious efforts, contact between New Mexico and the outside world did not cease. In 1816 a St. Louis trader by the name of Jules De Mun came to Taos to inquire whether trade with New Mexico might be possible. After forty-eight days in the calabozo, De Mun was sent packing, but not before he had time to note that the tributaries of the Rio Grande between Taos and Santa Fe "abounded with beaver." De Mun returned to St. Louis some $30,000 poorer by his own exaggerated count, but he took with him a knowledge of the riches to be found in New Mexico, and when the time was right, others like him would be back.

The right time came in 1821 when the first chapter in Mexico's long revolution ended, and the new nation declared its independence from Spain. The Mexican government quickly reversed the most stifling of the old colonial policies, among them the prohibition against trade across the northern frontier. Even as news of the change raced eastward, an American trading party from Missouri, led by William Becknell, was heading west in anticipation of it. Becknell reached Santa Fe on November 16, 1821, with seventeen men and a long string of pack mules. He set up shop on the plaza and immediately sold his calicos, hairpins, and hardware for a handsome profit, then turned and hurried back to Missouri to prepare an even larger cargo of merchandise for the next year. But thirteen of his men stayed behind, the majority of them electing to pass the winter in San Miguel del Vado, the eastern portal of New Mexico, which was an excellent place both to receive news from across the plains and to stay out of sight of the government officials in Santa Fe. Although the men's exact intentions are unknown, San Miguel would have made a good base from which to trap the Rio Pecos, including its headwater streams in the Sangres, where the cold mountain water caused the beavers' fur to grow rich and thick.

Commerce along the newly opened Santa Fe Trail and the sudden invasion of the mountains by fur trappers went hand in hand. Many of the men who helped work the trade caravans across the plains carried traps among their personal gear and stayed to try their fortunes in New Mexico. Ewing Young and William Wolfskill did so in 1822, having crossed the plains with Becknell's second expedition, and after a brief and unsuccessful attempt at manufacturing gunpowder, they repaired to the Upper Pecos, where they trapped beaver until the winter snows drove them back to the settlements in January. The traders who chose to return to Missouri, meanwhile, more often than not carried a cargo of pelts on their eastward journey. The furs made a handy currency in the early days of the Santa Fe trade since they were not subject to customs duties as specie was and since hard currency of any kind was in short supply in New Mexico.

Logically, the trappers first turned to the mountains closest at hand, and as a result, the Sangres between Santa Fe and Taos were the first mountains in the Southwest—and among the first in the entire West—to be ransacked for beaver. Besides Wolfskill, Young, and other Becknell men, a large trading and trapping party led by Thomas James began to operate out of Santa Fe in late 1821. James's partner, John McKnight, meanwhile, hurried south to Chihuahua to free his brother Robert, who had been jailed in 1812 for trying to open the Santa Fe trade nine years too soon. They arrived back in Santa Fe in early 1822, by which time a third American trapping party under Hugh Glenn had established itself in "Touse." To their surprise, however, Glenn and his men discovered that they were latecomers. At least one stream near Taos in which the expedition set its traps had already been trapped out, the beaver taken. Probably the depleted streams were the legacy of trappers who had wintered illegally in Taos before Mexican independence officially opened the way for commerce and trade. Later trappers, however, found worse. As 1821 drew to a close, the waters of the Sangre de Cristo Range were being exploited with lightning swiftness. The beaver in them did not last long.

Thanks to the efficiency of his technique, an experienced trapper could strip the beaver from a mountain stream in just a few days. The trap itself, which cost between twelve and sixteen dollars, was made of rough steel with big-toothed jaws that were sharp, but not too sharp, and powered by a spring that was strong, but not too strong—the idea was to capture the animal's leg, not amputate it. The trap was set in shallow water, and above it the trapper hung a green twig of aspen or willow, which he anointed with one or two drops of a vile liquid that he always carried with him and of which he nearly always stank. Part castorum (an extract of the beaver's scent glands), part gunpowder or spit or whatever other adulterant intuition or folklore suggested, this rank potion was the catalyst upon which the fur trade, and with it, most of the early exploration of the American West depended. It was irresistible to beavers. Scenting it, a beaver felt compelled to come closer, to read its message. But the perfumed twig was always a little too high; the doomed rodent had to get right under it, stand up on its hind legs, and ... SNAP! Since beavers are mainly nocturnal, this fatal drama was usually played at night. In the morning, provided he had set his trap well, the trapper would find the beaver drowned.

The pelts were fleshed, stretched, and dried. If he were far out in the wilds, a trapper might build a machine out of logs and poles with which to compress his skins into neat, dense bales which he could easily pack on his mules and haul to market. In the southern Sangre de Cristos, however, a man need not have gone to such trouble. Taos and Santa Fe were close enough that it made more sense simply to bundle the skins loosely, take a day or two to carry them in, and then return to the mountains for another round of stream walking, trap setting, and pelt curing.

Throughout the 1820s and 1830s Taos was the most important fur-trading center in the southern Rockies. Scores of trappers came there annually to sell their furs, provision themselves, rest, and certainly not least, to debauch and drink. In the latter regard two of the most important local resources were the handsome Taos women and the potent firewater, famous throughout the West, called "Taos lightning," which a group of enterprising Americans distilled from local grain. The names of the trappers and traders who based their operations in Taos make a roll call of the leading Anglo explorers of the Far Southwest and their exploits, which seemed as incredible to their contemporaries as they do now to us, were later made famous by dime novelists. William Wolfskill, among many other achievements, blazed new trails between New Mexico and California, and so did Isaac Slover, who arrived in New Mexico with Hugh Glenn in 1822. Ewing Young cap-

Photo 24. The Pecos River, in the heart of today's Pecos Wilderness, was once a favored trapping ground. Author photo.

Photo 25. Kit Carson, mountain man, explorer, Indian agent, and commanding officer of the 1st New Mexico volunteers. Courtesy Denver Public Library, Western History Department.

tained an epic expedition that lasted a year, covered a thousand miles or more of unknown territory, cost the lives of a third of the men involved, and returned to New Mexico with furs worth twenty thousand dollars. (Unfortunately for Young and his cohorts, the New Mexican governor promptly confiscated the furs.) Antoine Robidoux and Etienne Provost, working separately, opened much of the Green River country of eastern Utah, and Ceran St. Vrain, who commanded wide respect both from Anglos and Hispanos, became an important leader in the military, political, and economic affairs of the region. Most famous of all was Christopher "Kit" Carson, who while still a teenager served Wolfskill and Young as a camp cook and later became a national hero as a scout for the peripatetic "Pathfinder," Captain J. C. Frémont.

For every trapper remembered in history, there are scores of others whose stories remain unknown. The majority of the men who "cleaned" the beaver from the streams of the Pecos High Country and the rest of the southern Sangres belong to this category. One reason for their general anonymity was that they usually worked secretly in order to foil the attempts of the New Mexican government to regulate and tax them. Another was the simple geographical convenience of the mountains. Independent trappers, either alone or in small groups, operated easily in the Sangres. Unlike the large expeditions by which American trappers pressed westward into Arizona, the Great Basin, and California, these men left behind no wake of inventories, journals, and other records. So close were they to Santa Fe and Taos, they could work fast and unencumbered. They worked fast enough, in fact, that by 1825 when the Missouri *Advocate* belatedly reported that "north of Santa Fe the country is said to abound with beaver and several Missourians have returned richly rewarded for their labor," the entire southern half of the range had already been trapped out.

The winter of 1823–24 was probably the last good season of trapping in the southern Sangres, and undoubtedly a large number of the pelts exported to St. Louis the following summer came from the ravished streams of northern New Mexico. Augustus Storrs, a Santa Fe trader, reported to Senator Thomas Hart Benton that the value of that year's harvest amounted to $10,044, a considerable fortune by the standards of the day. Even so Storrs's figure was low, for, as he later observed to Governor Narbona of New Mexico, he had failed to include the haul of three other trading companies in his calculations. Assuming conservatively that the total value of the furs returned to St. Louis in 1824 was about $15,000 and that the price paid was between $4.50 and $5.00 per pound, the year's gross sales represent the skins of about two thousand animals—an enormous slaughter. Two thousand may not be a great many gnats or cabbages, but it is a tremendous lot of beaver. It may help to put the number in perspective to observe that in 1967, after decades of restoration and careful management, the population of beaver in the entire state of New Mexico was estimated to be no more than six thousand.

And year by year, the pressure on the furbearers kept increasing. Storrs noted that the number of trappers in Taos trebled between 1824 and 1825, and he predicted that the value of the furs shipped annually to St. Louis would pass $40,000.

The thoroughness of the trapping might not have awakened the soggy beaver in their streams to the possibility of extinction, but it did get the attention of the Mexican authorities. Responding to the growing scarcity of the animal, in 1838 the departmental junta of Chihuahua enacted one of the first conservation measures in the North American West by imposing a six-year moratorium on the trapping of beaver and otter in the Rio Grande. The ruling, however, was probably unenforceable, and certainly it came too late to have a significant effect. By then the furious energy of the fur trade had begun to wane. Styles in Europe and the United States were changing. Plummeting prices and the scarcity of beaver ensured that trappers would be poorly rewarded for their arduous and dangerous labor. Gradually they drifted into other lines of work.

The fur trade produced one of the most exotic and romantic episodes in North American history; it produced a body of geographic knowledge and lore upon which several decades of American conquest depended; it produced great fortunes, and a wealth of folklore; but one thing it did not produce was international amity and understanding. New Mexicans had initially welcomed the trappers and the St. Louis trading caravans that rumbled down the Santa Fe Trail. They had rejoiced at the prospect of obtaining more and cheaper goods than had ever been available from the sharp traders of Chihuahua. And many of the more farsighted among them had looked forward to an infusion of capital and technical know-how into the territory. Their expectations gave way, however, to a realization that the fractious and aggressive norteamericanos served no one's interests but their own.

New Mexicans could only watch as local streams were stripped of their riches with little profit to either the government or the people of the region. In the early years of the fur trade New Mexican authorities hoped to establish a domestic trapping industry and so issued permits for foreigners to trap so long as they took with them Mexicans who might learn the trade. But only a few of these apprentices ever became trappers in their own right; most simply functioned as servants. When Governor Narbona, in 1826, prohibited foreigners altogether from trapping, Americans adapted to the newly hostile situation in a number of ways. Some claimed that they had not trapped for their furs but bought them from Indians or Mexicans. Some hired Mexicans to obtain the necessary licenses and permits in their stead. Especially after 1830, many filed for and obtained Mexican citizenship—and then went into the business of concealing the actions of other Americans. But by far the greatest number resorted to the simplest tactic of all—smuggling. New Mexico's territory was too vast and its police and military forces too weak to prevent the Americans from buying, selling, and transporting their furs pretty much at will. Narbona complained, in fact, that he could spare only ten soldiers to guard the frontier at Taos but that they had no time to patrol for smugglers. Indian raids and the disorderly behavior of the Americans in town, he lamented, kept them fully occupied.

Like the fur trade, the more general mercantile traffic that Santa Fe traders called the Commerce of the Prairies also failed to live up to its original promise, at least from a Mexican point of view. Fairly quickly, it degenerated into a running commercial battle in which the traders and the governor's customs officers vigorously strove to outcheat each other. Tariffs were high and often arbitrary, bribery common, and smuggling rife. The stakes involved were greater than anything New Mexico had ever seen, for as soon as the American traders had saturated the Santa Fe markets with their muslins and mirrors, they turned their Pittsburgh wagons south to Chihuahua, Torreón, and other Mexican cities where they competed successfully with long-established Mexican merchants. Chihuahua, being a richer and larger city than bedraggled Santa Fe, soon eclipsed the New Mexican capital as the traders' primary destination. Many tens of thousands of dollars worth of goods annually passed through New Mexico en route to southern markets, but to the endless vexation of New Mexico customs officers, much of it never passed through Santa Fe. The Santa Fe traders typically stopped somewhere close to the Rio Pecos, packed their most valuable goods on mules, and sent them south on any of a number of wilderness trails. They then proceeded to Santa Fe and obligingly paid the 25 percent ad valorem duty on what remained of their freight.

New Mexicans profited from commerce with the United States on several levels. Some New Mexicans, for instance, became prosperous traders themselves. Nevertheless, many New Mexicans came more and more to resent the norteamericanos for the fortunes they arrogantly stole, for their swaggering air of cultural superiority, and for their sheer unmitigated rudeness. As he toured New Mexico in 1846, the English soldier of fortune George Frederick Ruxton had ample opportunity to observe the "bullying and overbearing demeanor" Americans showed New Mexicans, and he learned how it was received when he took lodging one night with a humble family in "Ohuaqui" (Pojoaque): "The *patrona* of the family seemed rather shy of me at first, until, in the course of conversation, she discovered that I was an Englishman. 'Gracias á Dios,' she exclaimed, 'a Christian will sleep with us tonight, and not an American!'"

The Americans, on the other hand, scorned the natives of the land for their race, their religion, their poverty, and for the military weakness that made flouting Mexican law so easy. There was not a printing press in the territory until one was finally hauled up from Mexico City in 1831, nor was there a college, a metal plow, or a sawmill. Everything about New Mexico seemed listless, moribund, and decadent. Even visitors from Old Mexico were shocked by the backwardness of life on the northern frontier. One of them, Antonio Barreiro, remarked that "crafts are in the worst state imaginable, even those which are indispensable for the prime necessities of life."

Still, the primitive conditions in which most New Mexicans lived do not account for the vicious attitudes that many Americans held toward them. After a visit to Taos in 1847, newspaperman Rufus B. Sage wrote: "There are no people on the continent, whether civilized or uncivilized with one or two exceptions, more miserable in condition or despicable in morale than the mongrel race inhabiting New Mexico." Inheritors of a strong cultural prejudice against all things Spanish and Catholic, most Americans were apt to find the New Mexicans "priest-ridden"—in spite of the fact that there were fewer than two dozen clergymen in the entire territory. And although they found plenty to praise in the brassy, short-skirted New Mexican women, who were opposite in every way to the prim Protestants back East, the Americans characterized New Mexican men as cruel and cowardly.

There were two main reasons for the antipathy between Hispanos and Anglos. One was the outright racism that the Anglos brought with them to the Southwest. Francis Parkman, the Bostonian chronicler of the Oregon Trail, expressed it as unabashedly as anyone. "The human race in this part of the world," he wrote, "is separated into three divisions, arranged in the order of their merits: white men, Indians, and Mexicans; to the latter of whom the honorable title of 'whites' is by no means conceded." The other and more subtle reason centered on the fact that the two cultures were at heart as different as oil and water. The Americans had begun to think that their national destiny was somehow "manifest" and that the entire continent beckoned them with an enormous economic and territorial mission. The New Mexicans, on the other hand, were essentially a pastoral people. Although their ancestors, the conquistadors, had been driven by a sense of destiny that was at the very least as aggressive as the Americans', the intervening years had bred in them an outlook much like that of the Pueblos: a settled feeling of permanence, a sense of completeness in the present, of being unchanged through time. Between cultures of such opposite temperament, conflict was inevitable.

7

Discord and Dollars

*Do not find it strange if there has been no
manifestation of joy and enthusiasm in seeing
this city occupied by your military forces. To us
the power of the Mexican republic is dead. No
matter what her condition, she was our mother.
What child will not shed abundant tears at the
tomb of his parents?*
      —*Juan Bautista Vigil y Alarid, acting
         governor, at the surrender of Santa
         Fe, 1846*

FOR REASONS THAT HAD as much to do with California and
Texas as New Mexico and that had more to do with national
spirit than with real estate, the United States went to war
with Mexico in 1846. Immediately upon the commencement
of hostilities, Colonel Stephen Watts Kearny set off down the
Santa Fe Trail at the head of the newly created Army of the
West. He intended to conquer all of northwest Mexico from
Santa Fe to Los Angeles.

Most of the major events of the turbulent coming months
in New Mexico occurred in the towns and cities surrounding
the southern Sangres: Las Vegas, Mora, Santa Cruz, Santa Fe,
and Taos. Only recently had permanent settlements on the
east side of the range begun to prosper. Pawnees laid waste to
Mora in the early 1830s, but the town, reoccupied for only a
decade, now consisted of two plazas several miles apart. Las
Vegas, named for its fine meadows, sprang up beside the Rio
Gallinas, where Santa Fe caravans liked to stop and rest before
embarking on the last arduous stage of their journey. Even as

Photo 26. Santa Cruz, New Mexico, c. 1882. Col. Sterling Price and Ceran St. Vrain planned their attack on the town from approximately this vantage in January 1847. The hottest fighting centered on the houses at right. Today this view is blocked by large cottonwoods and other riparian growth along the Santa Cruz River, which have become established as a result of reduced woodcutting and grazing pressure. Photo by William H. Jackson. Courtesy Museum of New Mexico.

the Army of the West approached, New Mexico's expansion had only begun. Its Hispanic people continued to spread outward into Indian country until the 1880s when the Hispanic vanguard of farmers and stock growers began to collide with the Anglo ranching frontier of the plains. Kearny's invasion of New Mexico in 1846 was but the first shock in a long series of collisions that continues to the present day.

The Army of the West included a small force of U.S. Army regulars, but most of its seventeen hundred men were volunteer Missourians, a hardy and spirited if ragged bunch. They marched into Las Vegas unopposed, and on the first morning of American occupation of New Mexico, Kearny stood on a rooftop and assured the local populace that he and his invaders would neither brand the New Mexican women on the cheek, as had been rumored, nor seize any farmer's crops or livestock without fair payment. The army continued toward Santa Fe, expecting a fight in the rugged canyons along the way, but took that city, too, without firing a shot. New Mexican military resistance, it turned out, had been entirely defused by the skillful maneuvering of James Magoffin, a Santa Fe trader who had traveled in advance of Kearny's army on a mission of intrigue commissioned by the president himself. Although it may never be known what exactly Magoffin did, it is probable that he purchased the nonbelligerence of the New Mexican governor, Manuel Armijo, with a large sum of his own money. The evidence for the bribe, admittedly, is slight, but after the war Senator Thomas Hart Benton of Missouri arranged for an appropriation to Magoffin of $50,000 later reduced to $30,000 "for secret services rendered."

Armijo, however, was not the only obstacle to a peaceful conquest. His principal military advisor, Colonel Diego Archuleta, also commanded the loyalty of the New Mexican militia and could have made an effective defense of Santa Fe. According to Benton, Archuleta "was of a different mould from the governor, and only accessible to a different class of considerations—those which addressed themselves to ambition." Magoffin apparently persuaded Archuleta that Kearny would take possession of New Mexico only as far as the Rio Grande, it being the western limit of a long-standing Texan claim to New Mexican territory, which was all that Kearny, supposedly, had come to assert. All the rest of New Mexico west of the river would therefore be Archuleta's to rule as he saw fit, provided he did not obstruct the American army in its march on Santa Fe. As gullible as he was ambitious, Archuleta raised no defense as the Missourians approached, and for his pains was swiftly double-crossed.

Kearny, now a general, took possession of all of New Mexico and set to work to secure it. He promulgated the necessary laws to form a territorial government and installed Charles Bent, an experienced but abrasive Taos trader and business partner of Ceran St. Vrain, as its head. Five weeks after taking Santa Fe, Kearny judged that the situation in New Mexico was sufficiently stable to allow him to press on toward California, leaving behind a small occupation force. That force subsequently became even smaller when the greater part of it was dispatched southward under the intrepid leadership of Colonel Alexander Doniphan to launch an invasion of Chihuahua.

The achievement of the Americans was astounding. An army of fewer than two thousand men, depending on a vulnerable supply line hundreds of miles long, had seized a huge and easily defended territory without the loss of a single life. The New Mexican militia, who were ready and eager to fight, outnumbered them by more than two to one and occupied superb positions near Pecos Canyon commanding the road to Santa Fe. Had the New Mexicans been led by men like Anza and Vargas, instead of Armijo and Archuleta, the outcome would have been very different.

Instead it was Kearny who emulated Vargas with a peaceful conquest of New Mexico, but also like Vargas, Kearny only postponed rather than eliminated violence. The tenuousness of the American grasp of New Mexico became abundantly clear on January 19, 1847, when Charles Bent, the Territorial Governor, was scalped and killed at his home in Taos by a loosely organized mob of Indians and Hispanos, who also put to death a half-dozen other Anglos and American sympathizers. A day later the rebels attacked Simeon Turley's distillery north of Taos in Arroyo Hondo, where famous Taos Lightning was made, and at the same time, across the mountains near Mora, Hispanos waylaid and killed a band of American traders on their way home to Missouri. Opposition to the American occupation spread rapidly, but it was poorly organized and miserably led.

Retribution came swiftly for the murder of Bent and the others. In Santa Fe Ceran St. Vrain and a contingent of former trappers and mountain men combined with Colonel Sterling Price's small force of regular troopers. Together the two groups, totaling 353 men, marched north, and at Santa Cruz they collided with a rag-tag army of New Mexican farmers, some fifteen hundred strong, who at first put up stiff resistance. The New Mexican force, however, finally collapsed under repeated frontal assaults by the outnumbered American infantry, and the Hispanos retreated in disarray, leaving behind thirty-six dead. U.S. casualties were two killed and six wounded.

Reinforced by Captain I. H. K. Burgwin, who had hurried north from Albuquerque with a company of heavy cavalry, the Americans continued to fight their way north. After a brief

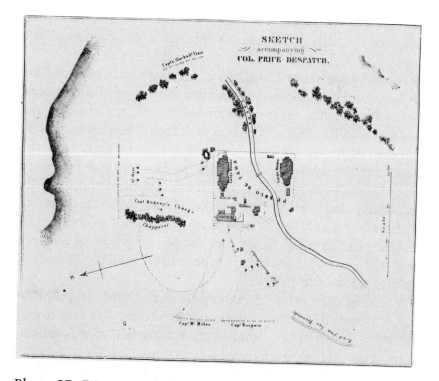

Photo 27. Diagram of the Battle of Taos Pueblo, as reported by Col. Price to the War Department and incorporated in Executive Document No. 8 of the 30th Congress, 1st Session. Courtesy of Davis Library, University of North Carolina.

engagement at Embudo they turned east into the mountains and marched through Las Trampas and Chamisal. They hauled their single cannon and a pair of small field howitzers across the snowy heights of the Picuris Range and descended, frost-bitten and weary, into Taos Valley. The last remnants of the insurgents, who included many Taos Indians, withdrew to the pueblo, where they made their stand inside the thick adobe walls of the church. For two days they held out, but finally the American artillery blasted a hole in the church big enough for a storming party and the defenders began to run. When evening fell, a hundred rebels lay dead amidst the ruins of the church and around the main compound of the pueblo. In the fields to the north lay the bodies of fifty more who had been cut down by St. Vrain's horsemen while trying to reach the safety of the hills. Price, meanwhile, counted seven dead and about fifty wounded, six of whom would later die.

The subjugation of Mora proceeded with only slightly less violence and bloodshed. A week after the killing of the traders, eighty Americans, most of them volunteers, marched on the town, which its residents rallied to defend. The ensuing battle raged from house to house around the main plaza and the American commander, Captain I. R. Hendly, and fifteen or more Morenos were slain, some in hand-to-hand combat. The defenders, however, prevailed and forced the Americans to withdraw to Las Vegas. There they complained to their commanders in Santa Fe: "If we had one or two pieces of artillery to scare them out of their dens, we could whip all the Mexicans on this side of the ridge."

A few days later the Americans did in fact procure a cannon and returned to Mora to make good their boast. But they found the town deserted, its inhabitants having fled to the mountains. The soldiers nonetheless determined to have their revenge. Training their cannon on the Mora courthouse, they commenced a day-long shelling that razed the town. The military confrontations in Mora and Taos were soon followed by a rash of executions in which various leaders of the insurgency, tried as murderers and traitors, not as Mexican patriots, were put to death.

Although mixed bands of Hispanos and Apaches and other Indians continued to raid American grazing parties on the plains, the hostilities of the Mexican War were essentially over. One fiery convulsion had passed, leaving a legacy of bitterness and enmity that would persist for generations. But soon New Mexico began to feel yet another, more subtle kind of force, which, however slow to work, was no less earthshaking. Hard on the heels of the American army came the conquering American dollar.

During the 1820s and 1830s Santa Fe traders had brought New Mexico into the economic orbit of the United States by selling a wide range of manufactured goods in the territory. Now, however, the U.S. Army further transformed the economy of the region by purchasing many of the goods that the territory produced. With hundreds of soldiers to maintain in garrisons scattered throughout the Southwest, the Army's shopping list was a long one. Many of its supplies came from Fort Union, on the edge of the plains.

Fort Union was well endowed. It lay in a valley with plenty of water, grazing, and an abundant supply of timber on the green ridges of the nearby Turkey Mountains. Its builder was Lieutenant Colonel Edwin V. Sumner, who had camped in the valley while serving Kearny in 1846. A flinty New Englander and one of the very few Yankee cavalry officers in the U.S. Army at that time, Sumner felt that the fort should be near the hostile Plains Indians and far from the treacherous and immoral New Mexicans. Previously the Army command had been in Santa Fe, which Sumner passionately characterized as "that sink of vice and extravagance." Dry plains, he reasoned, would be more conducive to military efficiency than wet saloons.

Fort Union's military mission was essentially to force New Mexico's many hostile Indian groups (Navajo, Apache, Ute, Comanche, and Kiowa) into leaving both New Mexico and its vulnerable and vital lifeline, the Santa Fe Trail, alone. Toward that end Sumner had built the fort near the junction of the two main forks of the Santa Fe Trail, the Mountain Branch

Photo 28. Ruins of Fort Union, c. 1962. Col. Sumner's original fort consisted of some thirty log buildings scattered through the area in foreground. They were succeeded by the starshaped earthworks, visible in lower right, that were thrown up in anticipation of Confederate attack in 1862. A larger rectangular complex, whose ruins dominate the area today, was built in 1863–69 to serve Fort Union's ultimate role as supply depot for the U.S. Army's southwestern operations. Courtesy Museum of New Mexico.

and the Cimarron Cut-Off. The fort's strategic location on the main highway of the region recommended it for a second important function, that of arsenal and military supply center for all of the vast New Mexico Territory, which then included Arizona and southern Colorado.

Army goods were hauled across the plains in great freight wagons, unloaded at the fort, divided, repacked, and then sent out again to one or another of the rapidly proliferating military posts scattered across the Southwest from the Pecos to the Colorado River. Food, naturally enough, was among the supplies distributed through Fort Union, for the quartermaster's depot had the responsibility of feeding thousands of men. And in the interests of economy the officers soon found that it was cheaper to buy many foods locally than to pay freight costs across the plains.

The foresighted Ceran St. Vrain, among others, recognized that an economic revolution was in the making. Early in the 1850s he moved from Taos to Mora and built a gristmill. By 1863 Mora was producing some sixty thousand bushels of wheat annually, and St. Vrain was grinding most of it and selling the flour to the Army. Samuel B. Watrous, who built a ranch not far from the fort at the confluence of the Mora and Sapello rivers, found the Army's appetite for beef to be equally profitable, and many other settlers in the area made good money supplying the fort with corn, oats, and vegetables. Las Vegas thrived on lucrative Army freighting contracts, and many men from the mountain villages encountered cash money for the first time when they hired out for wages as teamsters and wranglers. The flow of dollars touched even the area's poorest farmers, who rode out to the fort to hawk all kinds of supplies from beans to broom straw. Democratically, it also extended to the bars and brothels of Loma Parda, a village smorgasbord of carnal diversions, which contrary to Sumner's hopes had sprung up just across the ridge from the fort.

New Mexico was changing as never before in its history. Its population grew by almost 50 percent from 61,500 in 1850 to 93,500 a decade later, and over that time the number of Anglos in the territory increased threefold to perhaps as many as 3,000.

Some of those immigrant Anglos were Forty-Niners who came to New Mexico on their way to the gold fields of California and, losing their taste for travel, decided to go no further. Along with other newcomers, they saw in New Mexico the same things that Oñate and his colonists had seen: a land of irrigable valleys, infinite grass-covered ranges, and mountains that looked rich in minerals. The only barriers to taking up the challenge of exploitation and development were the American Civil War, which seriously affected New Mexico for only a brief time, and the "wild"—which is to say, non-Pueblo—Indians, who were discouragingly effective at checking the westward march of commerce. A correspondent to the Santa Fe *Weekly New Mexican* summed up the Anglo view of the territory when he wrote, "The rapid and irresistible current of events is fast approaching which must enlarge the importance of New Mexico and all her resources. With the Indians rendered powerless, we shall be able to profit by favorable events as they throng forward."

The troops of Fort Union were the servants of that vision, and their first major contribution toward its realization was to participate in an unfortunte and tragic war against the Jicarilla Apaches. Although the Jicarilla had long enjoyed friendly relations with New Mexico, U.S. military leaders were less willing, or less able, than their Hispanic predecessors to make distinctions between friendly and hostile nomads. That the Jicarilla were the first plains tribe to run afoul of the U.S. military in New Mexico was due mainly to the fact that they were hungrier than any other tribe. With game scarce in the mountains and competition keener than ever for the bison of the southern plains, the Jicarilla in 1854 were approaching starvation. In the spring of that year, a hundred warriors and their families camped near Picuris for the purpose of making earthenware vessels. The Jicarillas' micaceous cookware was popular among the Hispanic settlers of the region, and the Indians doubtless hoped to obtain food through barter and trade. When a small band of Jicarilla near Fort Union adopted a different strategy against hunger and stole some of Sam Watrous's cattle, Indian Agent Kit Carson and others began to fear that the peaceful Apaches at Picuris might be drawn into the Army's punishment of the thieves. On Carson's advice Acting Governor Messervy sent word to the Jicarilla that he would supply them with a weekly ration of corn or wheat if they remained at Picuris. Unfortunately, on the same day that Messervy sent his message, the Jicarilla struck their tents and moved out of Picuris. The commander of nearby Cantonment Burgwin dispatched a troop of sixty dragoons under First Lieutenant John W. Davidson to follow them.

The Apaches headed west along the Picuris Range toward the Rio Grande, joining that night with a Ute band of unknown size. Early the next morning, March 30, not far from the Hispanic hamlet of Cieneguilla (now Pilar), Davidson and his men made contact with them, and although it is not clear who attacked whom, the day eruped in bloodshed. It is not even clear how many casualties Davidson and his men suffered, but when the smoke finally cleared away, as few as twenty-two or as many as forty U.S. soldiers lay dead on the ground, making the battle of Cieneguilla one of the worst defeats the U.S. Army had suffered to that time in the western Indian wars. Summing up the furious but inconclusive pursuit of the Jicarilla that followed, Indian Agent Carson explained to the Territorial Governor that the Jicarilla

were driven into the war, by the action of the officers & troops in that quarter:—that since they have been attacked, with loss of lives, property, suspicions; vigorously pursued through the worst mountains I ever tracked through, covered with snow:—that their suffering & privations are now very great,—but that thinking there will be no quarter or mercy shown them, they will resort to all desperate expedients to escape any sort of pursuit & they have scattered now in all directions.

The Army never had its revenge against the Jicarilla, for neither the troops of New Mexican volunteers led by Ceran St. Vrain nor regulars from Fort Union ever caught up to them in their perpetual flight through the mountains. Instead the endless tracking of the Jicarilla gradually metamorphosed into a war against the tribe's more belligerent ally, the Utes, which lasted through 1855. During the subsequent Civil War years troops from Fort Union played a critical role in defending New Mexico and the gold-rich lands of Colorado and California from Confederate invasion, but even then they could not slacken their efforts against various hostile tribes that beset New Mexico. Sensing the vulnerability that resulted from the Civil War, a number of Indian groups stepped up their attacks on Santa Fe caravans and on settlements. The Fort Union command responded by launching a succession of year-long campaigns against, first, the Comanches and Kiowas, then the Mescalero Apaches, the Navajos, the Comanches again, and again the Mescaleros, and on and on.

War was an accepted constant of New Mexican life, and the Anglo military men of the new territory proceeded to wage it energetically. They were soon outraged to discover, however, that although all hostile tribes were essentially the same to them, the Hispanos and Pueblos of New Mexico were selective in whom they wished to destroy. Even while Comanches and Kiowas harassed the Santa Fe trail at great cost to life and property, New Mexican traders were able to roam the plains, locate the same groups of Indians that were successfully eluding the Army, and travel and trade with them peacefully. Later they would return to the settlements, their mules and carts heavily and profitably loaded with goods. Granted that there were Anglo traders farther north, notably William Bent, the brother of the slain governor, who engaged in the same dangerous commerce, but with a few exceptions there was no question about the loyalty of these men should hostilities grow serious. Not so for the New Mexicans. They were tractable neither to the short-term moods of official policy nor to the long-term appetite for genocide that the westering American people seemed collectively to possess.

The New Mexicans had been friendly with the Comanches (and with the Kiowa, with whom the Comanches were allied) ever since Anza's military prowess and diplomacy had paved the way for lasting peace between them. From that time forward New Mexican ciboleros and comancheros pressed progressively farther across the plains to follow both the great herds and the nomads who depended on them. Decorating themselves and their equipment with colorful pendants and banners, the New Mexicans proved to be remarkable plainsmen. With strings of pack mules and clumsy wooden-wheeled carretas, they freely roamed the full breadth and length of the giant tableland called the Llano Estacado or Staked Plains, which constitutes much of eastern New Mexico and the Texas Panhandle. The Llano is so dry that it has no rivers and so flat that rain, instead of running off it, collects to form shallow and exceedingly temporary lakes. In this forbiddingly monotone landscape, the Comanches enjoyed undisputed dominance over all their rivals.

The intrepid comancheros became as adept as their Comanche trading partners at navigating the trackless expanses

Photo 29. Jicarilla brave and his bride, newly married at Abiquiu Trading Post, 1874. Photo by Timothy O'Sullivan of the Wheeler Expedition. Stereopticon slides were an important and popular means for publicizing the work of the western surveys (see Ch. 11). Courtesy Museum of New Mexico.

of the Llano, and, also like the Comanches, they became skilled at avoiding the cavalry patrols of the U.S. Army, which earnestly mistrusted them and sought to restrict their trade. In the years following the Civil War the New Mexicans routinely passed along to the Plains tribes valuable tactical information about troop movements and the timing of campaigns. Moreover, the traders further aided their Indian friends by giving the military false and misleading information about the location of Indian groups.

For as long as the bison were plentiful, the comancheros traded for hides and meat, and when the bison grew scarce and the Comanches began to specialize in stealing longhorn cattle from Texas, the comancheros gladly traded for the cattle. They had no compunction about profiting from the misfortune of Texans. Texas had twice tried to conquer New Mexico, once in 1841 and once during the Civil War, and the contempt of Texans for "greasers" was well known in New Mexico and enthusiastically reciprocated. So unconcerned were the comancheros with the welfare of pioneers east of the Llano that some of them from time to time assisted Comanches in the actual theft of Texas cattle. Evidently there were also a number of Anglos and wealthy Hispanos in New Mexico who were equally sanguine about the Texans' losses, for according to New Mexican newspapers of the day, these entrepreneurs purchased the stolen herds from the comancheros and resold them, at considerable profit, in Colorado.

Although the comancheros were regarded as devils incarnate in Texas and excoriated in the Anglo-run New Mexico press, they were not much daunted by criticism. They gave no loyalty to the English-speaking nation that had engulfed them, only to their home villages and the rough frontier way of life that they had inherited along with their language and customs. They came from all over the Territory. Some were Pueblo Indians from the Rio Grande; some were genízaros from Pecos River settlements like Anton Chico, and still others were Hispanos from the old adobe plazas of the Sangre de Cristos.

To their own people the comancheros were heroes. Indian

trading was the get-rich-quick scheme of the day, and the goods they brought in from the plains eased life in their impoverished villages far more than any amount of farming could. According to Vicente Romero, a veteran of several expeditions to the Comanche, the successful return of a trading party occasioned "days filled with feasting and the nights with dancing. Blessed to God," he said, "those were the times."

Romero, a native of the tiny mountain village of Cordova between Truchas and Chimayo, went on his first comanchero expedition at the age of eighteen in 1868, the same year that the U.S. Army under Major General Philip Sheridan launched an all-out offensive against the Comanche-Kiowa alliance. With his uncle as comandante, Romero's band crossed the Sangres by way of Peñasco and Mora and then concealed itself as it neared Fort Union. If an Army patrol had discovered them, their planned three-month expedition would have been prematurely canceled, but they slipped by the fort under cover of darkness and made their way out to the plains.

Some trading parties met with the Comanches according to prearrangement at sites hidden in the caprock canyons along the Texas side of the Llano Estacado. Others, like Romero's, hunted for the trails of large groups in which there were signs of women and children. They followed the trail until it became fresh and then announced their presence with smoke signals.

The following morning the group would be surrounded by Comanches. A communal feast helped to instill in every one a spirit of commerce, and then the actual exchange of goods began. It was conducted as much by gambling as by outright trading. Heavy bets were laid on horse and foot races; men competed in bronc-riding and sharpshooting contests with bow and arrow. The comancheros had brought with them a variety of goods ranging from guns and munitions to cloth and copper kettles. Romero specifically mentions blankets (like those still woven today in Chimayo), strap iron for making arrowheads, and such foodstuffs as dried fruit, salt, and a hardtack-like bread that was baked especially for the Indian trade. He does not mention whiskey, but most comancheros, like their Anglo counterparts to the north, also carried a keg or two of rotgut with which to lubricate relations with their trading partners. The dispensing of the whiskey, however, was a delicate matter since, if ill-timed or ill-apportioned, it could lead to the riotous destruction of the trading party.

The Comanches, for their part, bartered with the resources of the plains: buffalo jerky, hides, stolen horses, mules, and cattle, whatever they had. Frequently they traded human beings, just as they had a century earlier at the Taos and Pecos trade fairs. Increasingly, through the middle of the nineteenth century, their captives were taken from the Anglo and Mexican settlements of Texas, and the fact that only the comancheros had the wherewithal to ransom these poor souls was considered by the rest of the world to be the sole redeeming aspect of their otherwise sordid trade. Not all the ransomed captives, however, were rescued so that they could be restored to their former homes. Romero mentions that in the course of one expedition of which he was comandante, an Indian offered José Antonio Vigil of Cundiyo two captive women (race and origin unspecified) in exchange for Vigil's excellent horse El Alazán. "This was a good offer, for these captive slaves were very much in demand among the *ricos* and prospective bridegrooms and brought a very good price."

Vigil, however, declined to part with his sorrel horse, and his refusal provoked an incident that nearly cost him and the rest of his party their lives. One Comanche, a chief called Capitán Corona, so desired the horse that he followed Vigil and his son out from camp one morning on an antelope hunt and, feigning friendship, rode along between them. Again he offered to trade for the animal, and Vigil did his best to humor him. Suddenly Corona clubbed Vigil's son with his left hand and knocked him off his horse. At the same moment he reached out with his right to unseat Vigil. Vigil, however, with lightning quickness and deadly resolve, drew his knife and plunged it into Corona's throat.

The Vigils buried the Indian in a shallow grave and crept back to camp after night had fallen. They told Romero to act as though Corona had taken them prisoner and to move on quickly before the Comanches found Corona's body. They, however, intended to leave immediately and make their way home as best they could.

Luckily for the comancheros a rain that night covered the tracks of the two Vigils, and although the Comanches were skeptical at first, they finally agreed with Romero that Corona had made captives of the two missing traders. And happily, Vigil and his son also survived their perilous trip back to the safety of the Sangres, but only after "many escapes from the Indians and from hunger."

Romero and his comrades were fortunate to conclude their trading peacefully; then, as was customary, their Comanche partners escorted them back toward the New Mexican side of the Llano. Again they slipped past Fort Union in the dead of night, and hurried on across the Sangres toward Cordova.

Their village lay huddled in a sheltered valley surrounded by scrubby tan hills. Upon reaching the ridge that overlooked it, the comancheros together fired their guns in a "Salvo to San Antonio," the patron saint of the village. At the sound of the shots, the people of the village hurriedly climbed to their rooftops, and from that vantage point, with great excitement and not a little dread, they studied the plainsmen descending from the ridge, and counted the living while anticipating the names of the dead.

8

Discipline

Adios, friends, for the last time
You have seen me wander round
Now lay me in my true home
In my coffin in the ground.
        —*Adios al Mundo, a*
         Penitente hymn

IN SPITE OF THE FAR-RANGING travels of Vicente Romero and his cohorts, the Cordova to which they returned, as well as all the other villages of northern New Mexico, belonged to a very small world. It was a world whose insularity, as well as its deep historical roots, accounted for the remarkable durability of Hispano culture. Unconsciously, indeed unavoidably, the Hispanic villagers of northern New Mexico resisted acculturation into the Anglo world much as the Pueblos had resisted the encroachments of Hispanos. Their extreme poverty shielded them from larger effects of the Anglo economy, for they had little money with which to buy the new things sold by Anglo traders. They had no schools and only scant interest in schooling, so the English language was slow to penetrate their mountain valleys. And they had no hunger for Anglo customs or ideas, for however poor they were in material things, they had a rich sense of their own authenticity. They were rooted in time, in tradition, and in place.

The world of Vicente Romero and his countrymen was a world in which barter, whether with nomads or neighbors, was the main form of commerce, and in which the business of agriculture began with yoking to a pair of gaunt oxen a homemade wooden plow, bereft of so much as a scrap of iron for the point of the share. Their houses contained no furniture except the rustic chests and cupboards the villagers made for themselves, no floors except packed dirt made hard with animal blood, and no windows except for occasional small openings protected by a sheet of mica or an animal skin scraped thin to a feeble translucence. The Hispanic world of northern New Mexico had been for so long—from its very founding, really—cut off from its roots in Spain and Mexico and so lacking in books and schools that its language had evolved into a peculiar, regional brand of Spanish, which Fray Domínguez, as early as 1776, had found full of archaisms and hard to understand. In other matters besides language the Hispanos of northern New Mexico had similarly wandered from the beaten path of their parent culture, and in no way did they do so more than in the practice of their religion.

Their relationship with God was a peculiar one, long lubricated with blood. They had fully inherited the Spanish fascination with morbidity and death, and in their mountain valleys they expressed that fascination as elaborately as it has ever been expressed anywhere. The name of the range—Sangre de Cristo—attests to their preoccupation, and so do the names of the two individual mountains of the high country, Penitente Peak and Hermit Peak.

In the years just before Vicente Romero's first trip to the plains, Hermit Peak was the home of an Italian named Giovanni Agostini, who lived in a small cave near the summit of the mountain. The reputation Agostini acquired for being able to draw water from dry rock and for curing the fatally ill spread throughout the mountain world and was nourished by the same love of mysticism and abnegation that sustained the already large and far more important Penitent Brotherhood. Through much of the nineteenth century, the Brotherhood, or as it was formally known, La Fraternidad Piadosa de Nuestro Padre Jesús Nazareno, was the most active religious force in the region, and in many quarters it possessed even greater authority than the official Catholic Church.

Photo 30. *Interior of a New Mexican home, 1912. Few Hispanos possessed any but the simplest furnishing until well into the present century. Photo by Jesse L. Nusbaum. Courtesy Museum of New Mexico.*

The origins of the Brotherhood have been much debated. Historians disagree whether the cult evolved slowly from another lay organization, the Third Order of St. Francis, which was present in New Mexico from the time of the Reconquest onward, or whether it was imported fully developed from some other part of Spanish America. Although the question of origins will probably never be resolved, there is general agreement that the Brotherhood emerged strong and ritually complete in the Santa Cruz Valley during the early years of the nineteenth century. At that time the distant Bishop of Durango, fifteen hundred miles to the south, was responsible for administering the colony's religious life, and the Church was so undermanned and so neglectful of New Mexico's outlying settlements that many villagers attended mass only once or twice a year.

Conditions had not improved by 1831 when Antonio Barreiro, a Mexican lawyer, came to Santa Fe and was scandalized by the state of the Church there. "Spiritual administration in New Mexico," he wrote,

is in a truly doleful condition. Nothing is more common than to see an infinite number of the sick die without confession or extreme unction. It is indeed unusual to see the eucharist administered to the sick. Corpses remain unburied for many days, and children are baptized at the cost of a thousand hardships. . . . How resentful might be the poor people who suffer such neglect!

How resentful indeed! Besides the neglect resulting from the shortage of clergy (including both Franciscan brothers and secular priests), New Mexicans also had to endure the unholy behavior of what few priests there were. Some of them lived in open concubinage and boasted of their children; virtually all were solicitous of the rich and neglectful of the poor; and they charged such high prices for their services that marriages and baptisms were frequently postponed for years or dispensed with altogether for want of money. The Church's dereliction became all the more acute as the Hispanic population of New Mexico almost trebled between 1810 and 1855 (to 50,000) and spread even farther from the core of the old colony near the Rio Grande pueblos. Before long, new settlements, far out of reach of the priests, were appearing in the San Luis Valley of Colorado, which was then still part of New Mexico, in the Navajo country to the west, and to the east in the caprock canyons of the high plains.

The Brotherhood expanded to fill the vacuum left by the Church, and every new village, as well as every old one, sooner or later possessed its own Penitente morada or meeting house. Depending upon the zeal with which Church authorities were condemning the Brotherhood at the time, the morada might be built in a secluded location where it would not be easily noticed, or, especially in remote communities, the morada might be in the center of town, as in Las Truchas, or even appended to the church as at Las Trampas and Vadito.

In the absence of a priest, the Penitentes kept alive the Church's yearly cycle of feasts and festivals with improvised observances of their own, many of which featured flagellation and other physical penances. Striving for a mystical oneness with the suffering Christ, they flayed their own and each other's backs with yucca whips, walked barefoot for miles over cactus and rough ground, and slaved for hours lugging wooden crosses as thick and heavy as railroad ties. Their bizarre activities reached a crescendo during Holy Week when vigils and mortification rituals and public processions ran on without a break, and in some years on Good Friday a specially chosen member of the village chapter was tied to a cross and hung on it until he lost consciousness, in a serious but nonfatal re-enactment of the crucifixion of Christ. Although a long and widely accepted folkloric tradition insists that many of these surrogate Christs did in fact die on the cross, there is little evidence to suggest that this was ever actually the case. The *idea* that death occasionally occurred, however, added considerably to the powerful mystique of the Brotherhood, as it was perceived both by outsiders and by the initiated.

The fascination of the Penitentes with death and physical suffering was unceasing. Nearly every chant and song harped on the theme of life's painful brevity, and most of the religious painting and sculpture produced in the villages of the northern mountains depicted saints either in agony or in morbid contemplation: St. Sebastian streaming blood from a dozen arrow wounds, St. Francis broodingly judgmental—with a skull in his hand. Most remarkable of all, however, was the frequent depiction of death itself: it was a doll carved out of cottonwood and given grimacing features. Dressed in black and sporting a black sombrero, it rode in Penitente processions in a vehicle of its own, a scaled-down version of the wooden-wheeled carreta that used to ply the trail to Chihuahua. According to legend, Death was sometimes an active participant in Penitente processions. Equipped with a tiny bow, which was always loaded and cocked, Death sometimes let fly its arrow as it was jostled along in its cart. Once loosed, say the storytellers, the arrow always scored a fatal human hit.

Penitente rituals, too, centered on the study of death. One of them, called Las Tinieblas, nowadays takes place on Holy Thursday night in a church, not a morada, and is open to the public. The church is dark except for a row of thirteen burning candles, one for each of the thirteen celebrants of the Last Supper, which are ranged along the altar rail. Following rounds of chanting and prayer, during which the souls of the recently deceased are commended to God, the candles, one by one, are extinguished, and the church grows steadily darker. It takes hours to work through all thirteen of the candles, but finally

Photo 31. "La Muerta." Courtesy Museum of New Mexico.

only one is left, and when it is put out, the church plunges into total darkness. Suddenly pandemonium breaks loose. A din of wailing, groaning, clanking of chains, and clacking of wooden noisemakers called *matracas* erupts, as the Penitentes create an energetic imitation of the dark chaos of perdition and Hell. As suddenly as it began, the racket stops, to be followed by more chants and prayers. But to the fearful delight of the children present, the howling and noisemaking break out several times more as the service and the night wear on.

As strange and gory as were some of the cult's practices, it is no wonder that the importance of the Brotherhood as a mutual aid society has often been overlooked. First and foremost, the Penitentes were a brotherhood, and each member swore to uphold the ideals of fraternity and community, as well as piety. Members took care of their brethren's funerals and looked after the families of those who were ill or who died. They softened the edge of village poverty by contributing to the support of their neediest neighbors, but in keeping with their desire to emulate the selflessness of Christ, these acts of generosity were typically anonymous and unpraised. The Brothers also may have functioned in a limited way as educators in their villages; sometimes the only education a young man received was his indoctrination into the beliefs and practices of the Brotherhood. Most important, however, the Brotherhood helped to unify the remote Hispanic communities of northern New Mexico and to reinforce the cultural pride of the villagers. New Mexican Hispanos needed all they could get of that kind of support, for the economic and social consequences of the Anglo conquest required extremely difficult adjustments.

Suddenly, with the advent of the Anglos, the Hispanos no longer stood at the top of New Mexico's social hierarchy, but occupied instead the middle place, or even the bottom. Many of the new Anglo elite who ran the affairs of the Territory considered the Pueblos, to whom Hispanos had always before had the luxury of feeling superior, to be more reliable, more trustworthy, and, by the standards of the day, more interesting than the Mexicans. The Indians had their share of vices, so the thinking went, but the racially mixed Hispanos had all the vices of the Indians and those of Spaniards, as well. The change in the social hierarchy was humiliating, and a villager was liable to sense it acutely when he went into town to trade.

The trading itself evoked difficult, conflicting emotions; there were so many opportunities to depart from the awkward, grinding slowness of the old ways, yet by accepting the Anglo way, an individual risked feeling dispossessed, disconnected. Now general stores sprang up everywhere, and sometimes a merchant from Las Vegas, Taos, or Santa Fe drove from village to village in a wagon with spoked wheels and iron tires. He brought new and necessary things: metal cooking pots to replace the old ceramic Jicarilla ware; tin cups, spoons, and forks; and iron grates that make cooking at the corner fireplace easier and faster. He also brought factory harnesses and stout iron chains that revolutionized farmwork; window glass, nails, and needles; all kinds of cloth; sturdy leather boots to replace moccasins; combs and patent medicines; coffee and sugar.

More and more a family needed to have cash money, and more and more it was impossible to avoid debt. The short-term solution to the money problem, which many villagers adopted, was to sell land, and in this they were eagerly assisted by a legion of Hispanic *ricos* and Anglo lawyers. In cases where villagers chose not to sell their land, particularly the valuable commons of their community land grants, the land was often gotten from them anyway, thanks in part to quirks in the new Anglo law that the rich and literate were able to exploit. The villagers had always been poor, but increasingly they were both poor and landless. The community spirit of the Brotherhood helped to assuage, for many, a troublesome sense of vulnerability.

The Penitentes were similar in many ways to other groups that have arisen in various cultures during times of severe and prolonged stress. Anthropologist Anthony F. C. Wallace calls such groups *revitalization movements* and defines them as "deliberate, organized attempts by some members of a society to construct a more satisfying culture by rapid acceptance of a pattern of multiple innovations." Wallace's model for the phenomenon is a little-known group of Seneca Indians who, beginning in 1799, followed the teachings of their charismatic leader Handsome Lake, but other, more familiar revitalization groups also come quickly to mind, including the Black Muslims and the followers of the Ghost Dance Religion. The Penitentes also fit Wallace's definition. The radical innovations they advocated consisted mainly of the new opportunities they offered for community organization and mutual aid, their emphasis on personal humility, compassion, and service to others, and their elaborate new rituals, which helped to unify the community by measuring the cycle of the year as, formerly, the calendar of the Church had done.

The Penitentes were a direct response to the convergence on the Hispanic community of three kinds of severe social change. First and historically earliest was the ecclesiastic abandonment that deprived the villagers not just of tradition and ritual, but of their main source of guidance and leadership. Second was the stress of rapid geographic expansion as new communities formed almost overnight while older communities suffered from the departure of the pioneers and the consequent breakup of families. And third was the psychic burden common to virtually every revitalization movement: a perception of lost status and cultural inferiority, which the Hispanos acquired as a result of the Anglo conquest.

It should be no surprise that by the 1880s probably a majority of the Hispanic males in New Mexico belonged to the Brotherhood. The Brotherhood may also have expanded in some communities to include women, although as a rule women supported the Penitentes only as *"auxiliadores,"* who helped with preparations for wakes and religious ceremonies, but were never initiated as full-fledged members. Many women, nevertheless, probably practiced various physical penances including wearing a "bracelet" of cactus under their clothing or crawling over rough ground to the *calvario,* or Calvary, where the Brothers enacted their ritual crucifixions.

Strong in numbers and united in outlook in the late nineteenth century, a number of local Penitente chapters began to dabble in politics. Predictably, their participation in public issues, although never entirely explicit, provoked angry condemnations from Anglos and Anglicized Hispanos. The reluctance of the Brothers to discuss their internal affairs publicly only added to the general fear that the Penitentes obediently submitted to the control of one demagogue after another. Their unseemly political character, as perceived by the rest of the country, was probably a significant factor in the delay of New Mexican statehood until 1912.

As it grew in strength and numbers, the Brotherhood also came into conflict with the Catholic Church in New Mexico, whose vitality had been restored following the consecration of Jean Baptiste Lamy in 1851 as the first bishop of Santa Fe. By anyone's standards, Lamy counts as one of the region's greatest and most humane benefactors. He established New Mexico's first college (which became the College of Santa Fe), the first hospital (St. Vincent's), and the first orphanage; he also struggled, albeit vainly, for the development of a system of public education, and with truly herculean energy he succeeded in infusing the decrepit Church with new life.

But Lamy was French, not Spanish, and in the wake of the Anglo conquest the Hispanos of New Mexico were not disposed to extend their loyalty or obedience to any foreigner. When Lamy and later his successor, J. B. Salpointe, who was also a Frenchman, undertook to curb the bloody excesses of the Penitente rites and to bring the brotherhood under the authority and control of the Church, the Penitentes resisted. They felt that they knew better what was necessary for the salvation of their Hispanic souls than did the French foreigners in Santa Fe. Ultimately, in 1888 Salpointe ordered his priests to withhold the sacraments from Penitentes who refused to mend their ways, but his attack on the order had little immediate effect. Rather than submit to the Church, the Brothers simply practiced their rituals in greater secrecy. In this way, as a religious underground, they continued for many years with undiminished strength.

Forbidden, they became even more fascinating to the outside world and excited the attention of a legion of popular

Photo 32. Penitente crucifixion. Photo by C. F. Lummis, 1888. While taking this picture, Lummis was protected by a bodyguard, Ireneo Chaves, who figures in the events of Chapter 12, along with his brother Amado. Courtesy Museum of New Mexico.

writers who vied with each other to bring this "little-known cult" to the attention of every magazine and newspaper reader in the country. One of the earliest and certainly the most intrepid of these publicists was Harvard-educated Charles Fletcher Lummis, whose book *The Land of Poco Tiempo* included the first photographs ever taken of an actual Penitente crucifixion. Lummis himself took the pictures in 1888, and he was assisted in his efforts by his friend Ireneo Chaves, who acted as bodyguard, threatening to shoot any Penitente who tried to interfere with the camera work.

The Anglos of New Mexico and the rest of the country were, of course, appalled by the Penitentes' extravagant practices. They recoiled at the crude self-punishments, which to them seemed barbaric and medieval. For their part, the Penitentes probably did not regret the Anglos' aversion. Anything that accentuated the differences between the fragile life of their villages and the sprawling, engulfing national culture was of service to the Hispanos' cultural survival.

After the turn of the century, however, the fortunes of the Brotherhood began slowly to decline. Undoubtedly the Church's persistent antagonism played some part in the gradual loss of membership, and so did the growing popularity of evangelical Protestant sects, but the principal cause for the decline was probably the increasing worldliness of the village people. Many of the men who left the villages for the sheep camps, mines, and railroad crews of the West, and who traveled even farther as doughboys in World War I, came to identify with the forward-looking spirit of the national culture and to drift away from the traditional attitudes that had sustained, and had been sustained by, the Penitente Brotherhood.

Surprisingly, the Penitentes did not die out. Instead they adapted to the new values by giving up their bloody extravagances, and in 1947 they finally won back the approval and endorsement of the Catholic Church. Today the Brotherhood still is active, though barely so, and possesses a membership of probably a little less than two thousand, most of it concentrated in the villages of the Sangre de Cristos. During Holy Week, Penitentes conduct processions and ceremonies in many communities, and although blood is no longer shed in public, some moderate physical penances are undoubtedly still practiced.

Today's Penitentes know that the largely non-Catholic Anglo society around them judges the Brotherhood to be an anachronism. But they do exist, and there is reason to hope that they will continue and even that they will flourish. The villages of northern New Mexico are today as embattled as they have ever been. They are economically depressed, plagued by crime and vandalism, and riven with factionalism. They need the spirit of personal discipline and community service that the Penitentes once helped to provide. Indeed, it may not be altogether coincidental that the loss of social stability within the villages and the decline of the Penitente Brotherhood have proceeded together.

The contemporary world scarcely has need of yet another cult, let alone one with as sanguinary a history as the Penitentes, but some individuals nonetheless find strong appeal in the religious ethos of the Brotherhood. They are attracted to the idea that physical pain, better than abstract dialogue, can make demons in the heart lie down. It is the appeal of an abiding belief in spiritual justice, which like many beliefs held by the villagers of the Sangres, expresses a deep fatalism. It is not so much a fatalism about future events as about the conditions of the present. One young Penitente expressed it well: "Some men," he said, "torture themselves all their lives with guilt in the pit of their stomachs. Other men deal with that guilt by doing something physical once a year. Either way, it is the same thing. You pay for what you get."

9

The View from the Top of the Cliff

Again the devil took him to a very high mountain and showed him all the kingdoms of the world in all their glory.

—Matt. 4:8

IF YOU DO GET what you pay for in penitential terms, then the hermit Giovanni Maria Agostini, who dwelled in a shallow cave on the mountain overlooking Las Vegas, now known as Hermit Peak, should have been rich indeed—in nonmaterial things. Although it is nearly impossible to separate the facts of his life from the folklore that has been created about him, there is no doubt that his life was as spare and ascetic as flesh and blood could bear. His shallow cave, a few hundred feet below the edge of a mountain cliff, was scarcely fit shelter for a pack rat. He obtained his water there drop by drop from the small seep that made his home perpetually damp, and he subsisted on an unvarying diet of atole, or cornmeal mush, of which he ate only enough to keep his body alive. So simple a life invited the curiosity of the villagers who lived in the hamlets below, and the stories they created about him reveal as much about the villagers as about the hermit.

Although there is no record of Agostini having involved himself with the Penitentes of the Hermit Peak area, his view of religion certainly had affinity with theirs. Like them he had little to do with Church orthodoxy or with priests, pre-

Photo 33. Hermit's Peak. Photo by Jesse L. Nusbaum, c. 1910. Courtesy Museum of New Mexico.

ferring instead to be his own guide along his path to God. Although he seems never to have discussed his relationship with the Church, Agostini, who had traveled the world and spoke several languages, may simply have been too independent to submit to any authority.

Agostini's one luxury was the extravagant beauty of his surroundings. His mountain, which formerly was known as El Cerro del Tecolote, or Owl Mountain, was a balcony over the planet. Below him, the cliff dropped away for a thousand feet. Before him spread the town of Las Vegas and the low rolling country that receded to the plains. The caravans of the Santa Fe Trail passed him in review, and their teamsters and passengers looked up at the hermit's brown, flat-faced mountain, glad to have reached the landmark of their journey's end, and fascinated as they crossed the threshold of strange New

Mexico, to hear and retell tales of the solitary man of the mountain.

Dawn has always been the best hour on Hermit Peak, and if Agostini did not rejoice in it, then indeed he was a madman. The top of the mountain, broad and flat, is clothed in a fine loose stand of bristlecone pine. The foliage of bristlecones is flecked with innumerable small beads of resin, and on Hermit Peak in the first light of morning the pine resin glitters like crystal; branches and then whole trees begin to glow like shining candelabra, and the grove fills with a soft incandescence. Swarms of chickadees and nuthatches flock to the light, flitting from tree to tree and dissolving the mountain hush in agitated chatter. To the east, beyond the edge of the cliff, there is no horizon, only the glare of the sun. The plains are on fire, bringing the morning from Texas. This is the dawn the eagles see, and this is also the dawn to which Giovanni Agostini awoke during the three and a half years that he lived on the mountain. One wonders what he thought of his eyrie. Before him spread such a smooth and golden sweep of space that at least once in a great while he must have felt the coals of his pride grow uncomfortably hot: here he was, a simple penitent, ascetic and devout, living high above the world of men, and close, he must have prayed, to God.

He had come to New Mexico in the summer of 1863, while the fortunes of the country pivoted on Gettysburg, having walked every step of the way from Council Grove, Kansas, 550 miles away. The captain of the wagon train in whose company he traveled was a member of the prominent Romero family for whom Romeroville, near Las Vegas, was later named. Wagonmaster Romero offered Agostini, already an old man, a seat in one of the wagons for the duration of the journey. Politely the offer was refused.

Agostini had introduced himself to Romero by showing him a sheaf of passports and letters of introduction collected from many years of almost constant travel in South America. Wherever he went, he had asked for, and nearly always received, the endorsement of influential people, including that of Emperor Pedro II of Brazil. These papers were still in his possession when he died in 1869, and although some have been lost, and the rest scattered among several owners, enough remain to provide at least some insight into the life he led before coming to New Mexico.

He was born in 1801 in the Piedmont village of Novara, Italy, the son of a nobleman, yet at the age of about thirty he renounced his patrimony and took up the life of an ascetic and a wanderer. Legend fleshes out those facts in various ways. According to one account, Agostini killed his cousin in a quarrel and gave the rest of his life to the expiation of that early sin. For nearly a decade he roamed over Spain, France, and Italy, making repeated pilgrimages to such holy sites as Rome and Compostela. In 1838 he sailed to Caracas, commencing the travels that would take him over the whole of the southern continent. His restlessness carried him to Pernambuco (now Recife) on the tropical Brazilian coast, to Tabatinga in the heart of the Amazon, to Motupe in the Andes, and far to the south to Patagonia. Everywhere he went, his behavior was strangely paradoxical. He sought out the rich and the powerful, whose cultured company he appears to have enjoyed, and then he withdrew from civilization to regions that offered solitude and separation from all but the simplest affairs of humankind.

The year 1859 found him on Mount Orizaba in central Mexico, living as he had throughout much of his wandering, in a cave. He had befriended the peones of the district, doctoring them when they were ill, giving counsel and comfort when they were in need. But in Mexico at that time it was a highly political, if not criminal, activity to show kindness toward the poor. The Mexican Rurales acknowledged Agostini's growing influence by deporting him to Cuba. From there he took a ship to Canada, and slowly he wandered southward to Council Grove and the Santa Fe Trail.

Agostini's first home in New Mexico was a cave in a canyon on the Romero family's huge ranch, but although he had come there seeking solitude above all else, he did not enjoy it for long. More and more as people learned of him, they took the path to his hideaway seeking prayers and cures. Finally in order to escape his own notoriety he moved to the mountain and made his home in the shallow, weather-beaten cave near the top of the cliff.

The hermit carved small trinkets and religious charms which he sold in Las Vegas for a few pennies each, and he used those pennies to purchase cornmeal, his only substantial expense. He may have tried to lead a life of the utmost simplicity and quiet, but stories of his ability to perform miracles nonetheless grew and spread. Inevitably the solitary, churchless Agostini was cast in the role of a saint, and he acquired a reputation for bringing the sick to health and for doing other things that suggested he was more than a mere mortal.

In the days when Agostini lived on the mountain, epidemics of smallpox and other diseases brought grief to every village not once but several times a generation. In 1856, for instance,

Photo 34. Hermit's Cave. Photo by Jesse L. Nusbaum, c. 1908. Courtesy Museum of New Mexico.

Mora Parish recorded 110 deaths from smallpox, nearly half of them infants. The only defenses at the people's disposal were herbs and prayer, and the hermit, who often descended from his eyrie to aid the sick, was held to be expert in the use of both. The broths and potions he prepared were credited with many acts of curing in those doctorless villages, and when there was no cure, his saintly manner at the bedside of the dying brought reassurance and relief. The priest was often distant, busy, perhaps expensive. The hermit Giovanni, known to the villagers as Juan, was nearer at hand, *muy simpático*, and he asked for nothing.

Agostini's most famous miracle is supposed to have occurred when a party of villagers from the nearby communities climbed the peak to build a cabin for him so that he would be better protected from the rigors of the mountain winter. Perhaps because of his age, Agostini consented. Trees were felled, logs hewed and notched, and slowly the walls of the cabin were raised. In keeping with Agostini's wishes, the cabin possessed no windows and only a single door so low that Agostini had to get down on his knees to pass through it. Some also say the doorposts bristled with sharp spikes, the better to try the flesh. Whatever the case, the day was hot, and the cabin's builders soon drank all the water they had brought with them. Unwilling to watch them suffer on his account, Agostini scratched the ground with his walking stick. Suddenly sweet water bubbled forth. Dry earth gave way to a miraculous spring.

The hermit's spring is a real spring, and today it is still the only good water on an otherwise dry mountaintop. And the hermit, whether he created the spring or not, was a real and ultimately mortal man. In 1867 at the age of sixty-six he left his mountain home and walked south, perhaps in search of greater solitude. Two years later his body was found in a cave in the Organ Mountains near Las Cruces, New Mexico, a dagger in its back. With that, the mystery of an unsolvable murder was added to the legends he left behind.

Today the waters of the miraculous spring are collected in a concrete cistern in a small tramped-down clearing about a quarter mile from the cliff. A wide, dusty trail leads as straight as an arrow from the cliff face to the spring, and until about 1979 it was lined all along the way with pairs of heavy wooden crosses, making a rustic, subalpine Via Crucia. Spaced at irregular intervals, the crosses were made of heavy, deeply notched logs joined to each other by finger-thick barn spikes, and they were fully big and strong enough to hang a man on.

They were built not, as many have alleged, by the Penitentes but by La Sociedad del Ermitaño, which formed in the years following Agostini's death for the purpose of keeping alive his example of piety and charity. The guiding light of this organization was Margarito Romero, scion of the same family that had befriended Agostini in 1863. Romero owned a great deal of land in the vicinity of the peak, and in 1890 he built a dude ranch at the foot of the mountain. He called it El Porvenir— the future. Actually the site of the ranch already had a perfectly good name, El Cerro de la Escondida, but Romero was afraid that English-speaking tourists would be unable either to pronounce or remember it. During the summer months a vacation at Romero's ranch cost seven dollars a week. Round-trip carriage fare from Las Vegas was a dollar more.

During Lent, however, there were no tourists, and Don Margarito put his ranch at the disposal of the Society, to serve as a staging ground for its annual pilgrimage to the peak. In 1908, at his own expense, he housed and fed nearly a hundred of his fellows on the nights before and after their climb. But following Don Margarito's death in 1917, the Society gradually disbanded. Today lone individuals still carry on the spirit of the group, but no one repairs the crosses on the trail to the spring or those that line the narrow trail that leads a few hundred yards down from the mountain top to the wind-blasted cave where the hermit once lived.

The main trail between the summit and the world below— the trail that the Society had used in its processions—leads through a steep cleft that cuts into the cliffs. Without doubt

Photo 35. Tourist party atop Hermit's Peak, c. 1898. Courtesy Museum of New Mexico.

it is as arduous and wearying as any trail in the mountain range. The last two miles are a snarl of switchbacks with the end never in sight; climbing them, an ordinary traveler feels like a beetle in a can.

On that trail one sometimes meets pilgrims to the mountain, and on one occasion I met two Hispanos from Las Vegas who said they were on their way to the hermit's spring. The older of the two appeared to be about sixty. He carried a walking staff in one hand; his other hand was a bandaged lump, its fingers lost within thick white wraps of gauze. The younger man, whose hair was just beginning to gray and whose sideburns curved down his cheeks almost to his moustache, carried a gallon plastic jug, the kind that milk is sometimes sold in. It was a hot day, but the jug was empty.

Only the younger man spoke much English. He was going to the hermit's spring to bring back water to his home. Did I know, he asked, about the spring? Its water has never been like ordinary water. The hermit, he said, gave it the power to cure.

Cure what?

Whatever needs curing. He gestured to his partner's bandaged hand.

I asked him if he came often to get the water. He said no, not often, a little water went a long way. But sometimes he

Photo 36. Religious procession, Santuario de Chimayo, New Mexico, c. 1917. The Santuario de Chimayo, one of New Mexico's most beautiful adobe chapels, is said to be built on a site where the earth itself has the power to heal ailments and infirmities of all kinds. The Santuario is the focus of scores of promesas every year, especially during Holy Week. Courtesy Museum of New Mexico.

came to the mountain for other reasons. Most recently he had been there to fulfill a *promesa*, a vow.

Any *promesa* is a serious agreement, but evidently this one was more serious than most. The man with the gallon jug had promised to climb the Hermit Peak trail barefooted.

He made the climb early in the spring when the snows had melted just enough to make the route passable. His son went with him to carry the boots he would wear on the trip down— or possibly put on if he became too injured to carry out his pledge. It was windy that day, and he and his son met no one else on the trail. The rocks were sharp and cut his feet. The miles seemed to stretch on unyieldingly. Near the top there were long patches of hard, crusted snow through which he had to break his way, leaving wet red tracks. But finally in spite of the rocks and the distance and the sawlike crust of the snow, he reached the spring. And sat there a long while, amazed his ordeal was over. He bathed his bleeding, half-frozen feet in the water. It was strange, he said, in no time they began to feel healthy and whole again.

It had been cold that day, he said solemnly. If it had not been for the cold and the numbness it had caused him, he could not have endured the pain.

I can see that, I said. The cold made you able to do it.

He smiled, then narrowed his eyes. His silence told me that I had responded wrongly, misunderstanding what he had said about the cold. It wasn't the cold he was talking about. It was some Thing behind the cold. I must have looked confused. He grinned, and offered me one of his cigarettes.

While he lived on the mountain, Agostini kept an agreement with Sam Watrous of La Junta (now Watrous, New Mexico) to make a signal fire at the edge of the cliff once every three or four nights to show that all was well. Seeing no fire, Watrous and others would have come in a hurry. The arrangement between the two men was well known in the region, but it was never called fully into effect. There was always a fire.

Today a visitor to the mountain still finds fire rings scattered along the exposed edges of the cliff. Perhaps they have something to do with honoring the Hermit, or at least with doing what has always been done. Either way, it is a nice idea. Fire on the mountain. Everything all right. The crosses rot in peace.

10

Gold in Them Hills

*That Hermit . . . he sure had lots of gold. He hid it
all away when he left, and it has never been
found. There's sure gold there but I've not found
it yet. I've got colors, though.*

—George Beatty

IN THE NOT-SO-HUMBLE opinion of George Beatty, hard-luck
prospector and lifelong windbag of the Pecos high country, the
important thing to be remembered about the hermit Agostini
was not his saintliness or his alleged supernatural powers; it
was his wealth. Along with plenty of other gold-fevered men,
Beatty believed that the hermit had struck a rich lode somewhere high in the mountains and that until some kind of
lunacy drove him southward, he had worked the vein little
by little and hoarded its gold. It only remained for a more
sensible man to rediscover the hermit's strike, and then life
in the mountain country would come to its inevitable, high-
rolling fruition.

George Beatty may have been crazy, but he was probably
no crazier than many other Anglos who came to the southern
Sangres. From the days of the Taos trappers virtually to the
present, the mountains of northern New Mexico (and not necessarily the towns or cities) have attracted a special breed of
Anglo—either insanely adventurous, reclusive, or, when it came
to prospecting, optimistic. George Beatty was all three.

He was well aware that gold had been struck in California

Photo 37. Prospectors on the trail in New Mexico, c. 1882. This was George Beatty's customary mode of travel: on foot with pack burro. Photo by George C. Bennett. Courtesy Museum of New Mexico.

and Colorado and silver in Nevada. New Mexico too had had a taste of bonanza. In 1866 a band of Utes rode down from the Sangres to trade at Fort Union, offering, in addition to skins and hides, some unusually soft dark rock for barter. It was copper ore, as pure as it is ever found in nature, and W. H. Kroenig and William H. Moore, who purchased their directions from the Utes, soon found the source of it on the slopes of 12,400-foot Baldy Peak, northeast of Taos. Even before their claim was assessed, they found something more: nuggets of gold in the gravel of one of the mountain's creeks. Soon miners rushed to Old Baldy, and within a couple of years thousands of them were busily panning, sluicing, and blasting away on every side of the mountain. They built a town and named it after the daughter of one of the surveyors who laid it out: Elizabethtown.

The miners soon shortened the name to E-town, but the town grew in every other way. By 1869 it boasted a population approaching seven thousand, a newspaper, five stores, two hotels, a drugstore, seven saloons, three dance halls, and a roster of outlaws that included the West's premier train robber, Black Jack Ketchum. E-town set an unenviable record for lawlessness when eight men died of gunshort wounds in a single twenty-four-hour period; in 1870 the anarchy in E-town finally compelled the governor to call out federal troops to restore order.

Before the boom ended in the mid-1870s, the placers and lodes above E-town yielded nearly four million dollars' worth of ore. While that was a paltry sum compared to other strikes in the West, it was a fabulous amount for New Mexico, and, most important, it promised that still more mineral wealth might soon be found. As New Mexicans fondly pointed out, the Territory had been "but indifferently prospected," and they boasted that E-town's riches came from the same mineralized formation that had made an Eldorado of Cripple Creek, Colorado. The boosters proclaimed with feverish certainty that, according to the best geological authorities, the formation ran unbroken through some four hundred miles of New Mexican mountains and shortly would give life to a dozen Cripple Creeks.

The lure of the money metals brought thousands of Anglos, millions of dollars of capital, and a new entrepreneurial spirit to New Mexico. It stimulated road building, attracted railroad investment, and hurried the organization of counties, towns, and mining districts. It was a soothing tonic for the shortcomings of a struggling, backward frontier territory, and a heady stimulant for thoughts about the future. Every New Mexican town in the 1870s and 1880s had a touch of gold fever, but probably none so much as Las Vegas, where George Beatty bought his drill bits, black powder, and other supplies. Already a major freighting and cattle center, Las Vegas became the busiest and richest town in the Territory in 1879 when the tracks of the Atchison, Topeka and Santa Fe Railroad finally reached it. As each train unloaded a new gaggle of aspiring tycoons at the siding, the certainty grew that sooner or later someone would step out of the crowd to become the next Sutter, the next Henry T. P. Comstock.

George Beatty figured that, if there were any justice, he would be the man. According to his own logic, the next big strike after E-town was bound to come in the Pecos high country, and with obsessive diligence he hunted for it for more than thirty years. Beatty searched the inner fastness of the range as no one else did, clambering from one timber-choked canyon to another, sinking prospect holes wherever the hard quartzites showed promise of a mineralized vein. He sought to uncover fresh new lodes, and he also hunted for the mountains' legendary treasures: the Hermit's Gold, the Lost Mine of the Padres, the Treasure of the Jealous Frenchmen, and others. In the course of his wanderings Beatty became probably the first Anglo to know the wild interior of the Pecos high country intimately, and he left more of a stamp on the nomenclature of the place than any man before or since. He built a cabin squarely in the heart of the wild country, on a grassy bench in a chain of golden meadows that he called his "ranch." Today those meadows are known collectively as "Beatty's," and "Beatty's Creek" runs through them, and a pair of "Beatty's Cabins," which have no relation to the rough structure George Beatty built, overlook them from the west. (One cabin belongs to the Forest Service, the other to the New Mexico Department of Game and Fish.) Beatty built his cabin of logs and gave it a sod roof and a puncheon floor. A few steps from his door the Rito del Padre and the Rio Pecos growled as they came together, and in every direction the mountains rolled back, silent and manless and full of life. Beatty had chosen his site with a hermit's eye. Like any man who turns his back on settled life to live in the wild, he gave himself both to the perfections of nature and to the imperfections of his own unruly mind.

He dressed head to foot in fringed buckskins and cut a handsome figure in a hawk-faced, wiry way. There was a hint of humor in his thin, seemingly moth-ravaged goatee and moustache, but it was not sustained by his eyes. They had an agitated look. In the company of others, whether strangers or friends, Beatty always talked too loudly and too long. He was, they said, a character. No one knew him well.

Beatty was on hand in the early 1870s when traces of gold were discovered in the rough hills that roll south from Elk Mountain. Within days clusters of tents appeared in the meadows of Tecolote Creek and on the upper Rio Gallinas, and corporations were chartered on barrel tables by the light of kerosene lanterns. The atmosphere was as raw and heady as the whiskey of the camps. As soon as one strike proved worthless, another was made only a mile or two away, and the miners swarmed there. Mining machinery was ordered from Massachusetts, and at one point no less than thirty-eight wagonloads of drills, hoist engines, and other gear were hauled up into the Gallinas hills. But as happened countless times elsewhere in the West, the venture failed. The ores never "panned out" and would not justify the effort of mining.

Soon, however, strikes of gold, silver, and copper were scored near Rociada, and the throng—Beatty no doubt among them—rushed around to the north side of Hermit's Peak. Investments, if anything, increased, but although a few mines edged close enough to profitability to stay in operation a number of years, the Rociada ores also proved too poor to bear the cost of freighting to Las Vegas.

[151]

Mining failures far outnumbered the successes in the southern Sangres, and the savings of many investors—most of them from the East—bought no more than a pile of tailings and an abandoned shaft. In addition to their meager wages the miners themselves got all they could take of heavy, dirty, dangerous labor, bad food, and bad weather. But if a miner chose to be a prospector and work for himself, as Beatty did, he could trade some of the labor and all of the wages for solitude, independence, and rootlessness.

Rociada was Beatty's home as much as any place was. Nestled hard against the upthrust wall of the east divide of the southern Sangres, and with a broad green apron of pastures and croplands spread around it, the village had the look of a dew-laden Shangri-la. That is what its lilting, sibilant name means: bedewed. But the charms of settled life in Rociada were lost on George Beatty. He spent only his winters there, periodically working a few nearby claims, which like everything else he did, kept him in a state of perpetual impoverishment. In spring as soon as the snow melted enough, he would load his pack burro and head for the high country. The trails were few and poor in those days, and usually Beatty went on foot, hacking a path through the brush and timber with one of his "sure good bear knives."

He might go first to H. S. Harvey's dude ranch on El Cielo Mountain, where he terrorized the early season guests with bear stories and brandished his long knives in passionate re-enactment of his allegedly numerous "scrapes" with grizzlies. Beatty's formidable knives were more like short swords. He had hammered them out of mining drill bits, and their blades were easily a foot and a half long. Beatty said they were better protection against grizzlies than any gun that had been made. He claimed that when a bear attacked him he quickly lay down on his back with the knives pointed up. Thus prepared, he would instantly disembowel any bruin that leaped upon him. Smart bears, he said, always ran away. Whether or not Harvey's guests believed his preposterous stories, they must have found Beatty quite incredible in his own right.

From Harvey's he might push on to Elk Mountain, where

Photo 38. George Beatty with his "sure good bear knives,"
c. 1884. Photo probably by Prof. L. L. Dyche of the
University of Kansas.

he had a small cabin and a mica claim that produced a little income, and from there continue to his "ranch" on the Pecos. If hunting proved good, he might stay in the mountains for months at a time, living off the land and enjoying a solitary and probably cantankerous peace with his surroundings. He would poke with pick, shovel, and pry-bar at the canyon walls and cliff sides, laboring over his prospect holes with a religious kind of indolence and a singular lack of success.

In spite of nearly forty years of prospecting, George Beatty never found his Big Strike. As age shortened the radius of his wanderings, he was compelled to abandon his mountain haunts, and by the early 1890s the roof of his deserted Pecos cabin had fallen in. Nevertheless, Beatty never lost his manic prospector's confidence. He called frequently at the isolated ranches at the foot of the mountains and always accepted the obligatory invitation to supper. Men who were children on those homesteads can remember today how Beatty charged the evenings with energy as he proclaimed the virtues of his latest prospect. Only a foot more, or two at the most, and then he would hit pay dirt for sure. Beatty promised that for the insignificant cost of a small grubstake, he would reserve for his host a fat share of his imminently enormous wealth.

Harry Mosimann, who grew up on a ranch at the head of Sapello Creek, recalls George Beatty as a loud, strutting eccentric, well into his sixties, who called regularly at his family's homestead. Harry Mosimann's father liked George Beatty, and as old age and poverty made the prospector's life increasingly grim, the elder Mosimann did his best to provide Beatty with work. There was a round stone building on the ranch that had been gutted by fire, and about 1905 Mosimann hired Beatty to refit it with doors and windows. Harry Mosimann recalls that Beatty possessed considerable skill as a carpenter and that his work on the building was craftsmanlike, except for one thing: Beatty refused to work with a board that had a knot in it. No one could be entirely sure whether it was because he was paid by the day or because he was slave to some perversely high standard, but Beatty insisted on cutting the knots from every board he used and replacing them with carefully fitted plugs of unblemished, straight-grained wood. A time-consuming task, to say the least, but it was Beatty's way. Take it, said the crusty ex–grizzly fighter, or leave it. Generously, the Mosimanns took it.

After his death, George Beatty's expectation that the Pecos high country would yield vast quantities of mineral wealth was finally realized, but only with the help of a great deal of capital, machinery, and manpower. Beatty's day was over. A new era had dawned in which organization, planning, and technical innovation led most reliably to success, and mining was no exception.

The claim that ultimately brought modern industrial mining to the southern Sangres was first worked in 1882 by a group of men who organized themselves as the Pecos River Mining Company. Their mine, affectionately named Evangeline, lay a dozen miles downriver from Beatty's Pecos cabin, at a place that came later to be called Terrero. (*Terrero*, appropriately, is the Spanish word for mine dumps or tailings.) Evangeline penetrated a vein of fairly high quality copper laced with gold and silver. Unfortunately, there was so much zinc mixed in with the money metals that the smelters in El Paso, to which the ores were hauled, sometimes rejected them altogether. As a result, the mine operated only intermittently and made no one rich.

Evangeline passed into the hands of the Goodrich-Lockhart Company in 1916 and to the American Metal Company in 1925. American Metal (now known as AMAX and still a leader in hard-rock mining in the United States) gambled a tremendous amount of capital on the Terrero operation, and the gamble paid off. Over the next thirteen years the company extracted more than thirty-seven million dollars' worth of ore from the Precambrian formations beneath Terrero, and for twelve of those years the mine was New Mexico's leading producer of zinc, lead, gold, and silver.

During its boom years, the mine employed some six hundred workers and reached a total depth of 1,750 feet. It consumed a staggering six million board feet of timber for shoring *per year* and relied upon what was then the longest aerial tramway in North America to transport its ore in half-ton buckets from the mine to a crushing mill twelve miles away.

As long as the mine remained healthy, Terrero flourished. It boasted a population of twenty-five hundred, a two-chair barbershop, a twelve-bed hospital, an elementary school, and a red-light district called Chihuahua. It was on the itinerary of every small-time prizefighter in the Southwest, and year after year it yielded one of the best baseball teams in the state. Those were the days of Babe Ruth and Lou Gehrig, and the American Metal Company took the game seriously. Its players were given jobs as "muckers," which was the only position at the mine exempt from the daily work quotas that ruled the lives of other miners. Although dangerous, mining work had advantages, for life above ground was full of both public and

Photo 39. Terrero in 1935. Today virtually none of these buildings remains. Photo by T. Harmon Parkhurst. Courtesy Museum of New Mexico.

private diversions. According to some veterans of those days, it was not unusual for a miner to have one wife back home in Rociada or Mora and another with whom he lived in Terrero.

The Great Depression, however, brought hard times to the mine and to the underpaid and overworked miners. Labor unrest culminated in a three-month strike in 1936, by which time day-to-day operations had become extraordinarily difficult on account of underground flooding. Fourteen tons of water had to be pumped out from the shafts for every muddy ton of ore that was hauled up. Together, labor demands and engineering difficulties caused the mine to shut down in 1939, and with that Terrero became a ghost town. Today only a bar and a grocery store remain near the site of the once wild and woolly mining town, but the mine still makes its presence felt in a strangely eerie and unsettling way. During spring melt and following heavy thunderstorms people in the area sometimes feel a low vibration underfoot and hear a ponderous rumble. It cannot be thunder because it so often occurs on cloudless, stormless days. What it is, some insist, is the muted crash of empty tunnels collapsing in the dark Precambrian infinities below.

In spite of the great quantity of treasure that American Metal hauled from the Terrero mine, a far greater amount may yet lie hidden in the mountain rock, waiting to be let out. Two groups seem to have faith in the possible existence of those riches, and one of them consists of a half-dozen or so mining corporations, most of them subsidiaries of major oil companies, who since the mid-1970s have been busily evaluating the potential of the southern Sangres for the profitable production of copper, gold, silver, and other metals, especially uranium. Whether development of any mineral reserves takes place will depend not only on market forces and the quality of the ores but on the willingness of the Forest Service to permit economic development of the land immediately surrounding but not included in the Pecos Wilderness. Any decision by Congress to make wilderness areas more accessible to mining and mineral development would have even greater impact. With so many variables involved, the future of mining in the southern Sangres is impossible to predict, but certainly if significant development should take place, it would dramatically alter both the depressed regional economy and the relatively pristine mountain environment.

The second group with an interest in the mountains is far smaller and less powerful, but no less persistent. It includes the psychic heirs of George Beatty—and for that matter, of Coronado, Oñate, and all the other hopeful prospectors, great and small, who have chased their dreams across the Land of Enchantment. For over four hundred years men have repeatedly and energetically made outlandish claims for New Mexico. At about the same time that George Beatty was plugging the knotholes in the Mosimanns' window frames, Fayette Jones, author of the influential book *New Mexico Mines and Minerals*, intoxicated a generation of placer prospectors with the outrageous boast that "the whole channel of the Galisteo river is one vast sluice box; the gold collecting in the depressions and seams of the rocks that form the river bed. The same thing will apply to the Rio Grande. . . . The trough of the Rio Grande is a vast storehouse of gold."

Besides the gold scattered in the creekbeds and the lodes encased in quartzite and amphibolite that remain to be found, there are also the treasures, already packed up and ready to haul away, that the miners of the past have lost or hidden in the high country. No one has yet found the Hermit's Gold that Beatty hunted, although people continue to try. And the lore of other buried treasures and lost mines supposed to exist near the headwaters of the Pecos has already filled several chapters in a number of books. Arthur Campa tells the story of the Lost Mine of La Jicarita in *Treasure of the Sangre de Cristos*. In *Coronado's Children*, J. Frank Dobie includes a half-dozen tales of enchanted mines that he collected from an active treasure hunter in Pecos in the 1920s. And Elliot Barker, in *Beatty's Cabin*, muses on the legend of the mysterious and anonymous Padres, whose lost mine is supposed to lie somewhere near their namesake, the Cerrito del Padre.

My favorite among the old tales, however, is the story of the Treasure of the Jealous Frenchmen, which appears to be rooted in real events. The germ of the story was the unhappy experience of a band of French trappers, outfitted by Charles Beaubien, who set off to try their luck in the mountains above Mora in 1838. Evidently they anticipated that the beaver of the southern Sangres had by then recovered form the devastation they suffered a decade earlier during the heyday of the fur trade, but the Frenchmen never brought back a single pelt. Every one of them was killed by Indians—or so concluded the men who discovered their bodies and the clerk at Beaubien's store who recorded their misfortune in his journal.

Over the next hundred years, however, the villagers of Mora and the rest of the mountain country concluded otherwise. They elaborated the story of the Frenchmen's death into a tale of obsessive greed and Indian sorcery that harkened back to the earliest days of Spanish settlement. Early on, so the story went, the Pueblos had realized how much the Spanish coveted gold, and in order to frustrate this avarice, their medicine men had placed a curse upon it.

It was this curse that ruined the Jealous Frenchmen, for although they found gold in the mountains and hauled enough back to civilization to fund years of carousing and fast living, when they returned from France to obtain a second helping, their lust for riches began to overtake them.

On the voyage across the Atlantic, two of the Frenchmen opted for a less complicated division of the spoils and conspired to throw the third partner overboard. These two then worked the mine together for a short time and returned, as before, to civilization to consume their riches in wild pleasures. A third trip to the mine followed, but from it only one Frenchman returned to enjoy the accustomed round of decadence alone. Predictably, when his supply of gold was exhausted, this two-time murderer made a fourth trip to the mine, which was his last. A few days after his expected return, his burro wandered alone into Mora, and soon afterward a sheepherder found the man's lifeless body. On its forehead was a red mark in the shape of a tiny hand, and beside the body lay a leather sack containing, instead of gold, a collection of ordinary and worthless black rocks.

The story of the Jealous Frenchmen lives on. Not many years ago, I met a man named Whitey, who with the aid of an understandably disgruntled teenager, was lugging a heavy metal detector through the highest reaches of the Pecos Wilderness. He confided that he possessed a map showing the exact location of the treasure of the Jealous Frenchmen, and to my amazement he allowed me to see it. With great precision and confidence he told me that the treasure consisted of bullion buried twenty-two feet beneath the ground. Unfortunately, bad weather and confusing trails prevented him from finding it on this particular expedition. He would come back next year, he said, and he would improve his preparations by bringing a pack horse instead of a teenager.

I am sure that Whitey did come back, and I am also sure that many other latter-day George Beattys still hurry through the mountains on the track of wild dreams. In spite of the pragmatic diligence with which the Forest Service manages the mountains' inert resources, they will remain for many a land of mystery and a frontier of personal adventure, as once they were the frontier of whole cultures. From Beatty's day forward, however, the future of the mountains lay not with solitary adventurers but with essentially conventional individuals whose lives expressed the best and the worst of Anglo society.

BOOK TWO

COLLECTIVITY

11

Inventory

*To a great extent, the redemption of all these
lands will require extensive and comprehensive
plans, for the execution of which aggregated
capital or cooperative labor will be necessary.*
　　　　　　　　　　　　　—John Wesley Powell

THE UNITED STATES CONQUERED New Mexico twice. The first conquest was carried out by traders, miners, ranchers, and speculators—"rugged individualists" who served only themselves. The second conquest proceeded concurrently but was the work of soldiers, scientists, and other professionals who represented the United States as a collectivity. Their efforts toward protecting the interests of the public as a whole culminated in the creation of such institutions as the United States Geological Survey, the National Park Service, and the Forest Service. The conflict between these two very different conquests sparked acrimonious debate and even gunplay in every state and territory in the West, but nowhere was it more complex or intense than in New Mexico, where Hispanos and Pueblos, possessing other collective interests, struggled to shape their own future.

The collective and scientific American conquest of northern New Mexico passed a milestone of sorts in the middle of the 1870s when the first representatives of the U.S. government climbed into the high country of the southern Sangres. At that time George Beatty had just finished building his solitary cabin on the Upper Pecos, and the mountains remained largely unexplored. They were still wilderness in the true danger-laden sense of the word. Only a few years earlier a band of outlaw Jicarilla, worn out from fighting Texas Rangers and U.S. Cavalry on the plains, had decided to trade the perils of the warpath for frostbite and near starvation by making their winter camp near the headwaters of the Rio Pecos. Beatty may have crossed their trail as he also crossed the trails of wandering sheepherders and hunters. The mountains were big enough in those days to absorb a fair number of men and yet hold them apart one from another, like a few scattered fleas on a very large dog. And so it is not likely, though neither is it impossible, that Beatty, who epitomized the individualistic, mountain-man past of the high country, may have stood for a moment face to face with the army surveyors Blunt and Maxson, the first wave of the scientific and federalized future.

First Lieutenant Stanhope E. Blunt and his assistant topographer, F. O. Maxson, a civilian, penetrated the wilderness at the head of the Pecos in October 1874. They were accompanied by a cook, a wrangler, and two packers, and also by a second topographer, a meteorologist, and an additional individual whose difficult job was to tend the temperamental, wheeled odometer that measured the length of their travels. They had come to the mountains in order to make maps, an almost entirely new enterprise in the Sangres. Whereas Beatty and his fur-trapping predecessors had kept their intimate knowledge of the mountains private and personal, Blunt and his party were committed to making knowledge of the forests and canyons a public possession. They were one of several field expeditions belonging to the comprehensive "United States Geographical Surveys West of the One Hundredth Meridian," which was the Army's final grand attempt to complete the exploration and mapping of the West.

Commanding the surveys was an ambitious young first lieutenant named George Montague Wheeler who had graduated from West Point in 1866 near the top of his class and proposed the idea of the survey a few years later. He immediately received strong support. The Army had fallen on dull and unglorious days after the Civil War. Avenues of promotion were clogged in the overgrown officer corps, and morale was dangerously low. Wheeler's argument for a new and vigorous western adventure promised to restore some of the luster of the military's wartime reputation.

After he proved his effectiveness in a long and difficult reconnaissance of eastern Nevada in 1869, Wheeler at last received authority to initiate his plan. What he contemplated was an encyclopedic exploration that would compile for the Army's use all pertinent "astronomical, geodetic, and topographical observations, with map delineations of all natural objects, means of communication, artificial and economic features"—that is, all the information a military strategist might ever need. Wheeler's hope, which was enthusiastically shared by his superiors, was to return the Army to its former preeminence as pathfinder and discoverer of the West, a position increasingly usurped by the scientifically oriented civilian surveys of Clarence King, F. V. Hayden, and John Wesley Powell.

Wheeler's foremost mission was to collect military intelligence, but he could ill afford to neglect the natural history of the provinces he mapped. The West in those first post-Darwin decades had come to be regarded as a vast laboratory of the natural sciences, and major discoveries in geology, paleontology, and other disciplines were harvested there like rich wild crops. The more his competitors, Hayden and Powell, accused him of being dull-witted and unscholarly, the more Wheeler and his men were obliged to champion the causes of science, as well as those of the military.

Two of the Wheeler Survey's most impressive scientific coups were struck in New Mexico. One was Professor E. D. Cope's discovery of extensive Eocene fossil remains in the dry Comanche country of the northeast plains, and the other was Cope's joint exploration with Dr. Oscar Loew of Anasazi ruins in the northwest of the Territory. Although less dramatic, the survey's scientific investigations of the Pecos country were pioneering and thorough. Cope published a report on the area's paleontology in 1877, the same year that Loew added his observations on its geology to those published by A. R. Conkling

Photo 40. Wheeler Survey map of north central New Mexico. Considering the equipment of Wheeler's day, the elevations given are very good, being generally within forty or fifty feet of those accepted today, but in other respects the map is inaccurate. It reduces the breadth of the upper Pecos watershed from thirteen to nine miles, an error of nearly 30 percent, and it omits some of the range's most important streams while showing others collecting their headwaters from nonexistent peaks. These failings may be uncharacteristic of the overall excellence of the Survey's work, but they nonetheless lend credence to criticisms voiced by members of the Powell and Hayden surveys that a certain amount of Wheeler's topographical work was "left to the imagination of the draughtsman." Accuracy aside, the Wheeler maps were unsurpassed artistically and represent the height of map-making achievement in the years prior to the adoption of contour lines as the preferred means for depicting topographical features. Courtesy Alex Harris.

in 1876. Additionally, every survey party that explored the mountains kept detailed records of the birds and other fauna they encountered, even to the point of cataloguing the Sangres' alpine insect life.

Science notwithstanding, the most important mission of the Wheeler Survey in the Sangres was topographic mapping. The high peaks provided the essential observation stations from which the main physical features of the surrounding plains and valleys could be related one to another by means of triangulation. Although the highest peak in the range is named for him, Wheeler himself never came into the Sangres. He devoted his efforts to the more difficult and newsworthy work of ex-

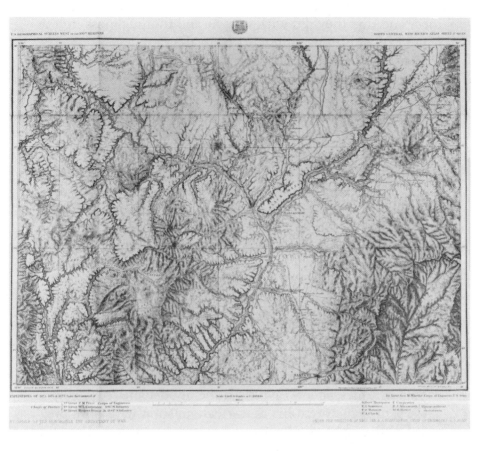

ploring the Colorado River and the hostile deserts of the Great Basin. But Stanhope Blunt carried his transit to the top of most of the Pecos country's tallest mountains in 1874, and any mountains he missed were climbed a year later when F. O. Maxson returned to the mountains in temporary command of his own crew.

No less than solitary George Beatty, the soldier-scientists of the Wheeler Survey came to the mountains in search of treasure. All the westward expansion of the nation was a great, collective treasure hunt, and for a short time Wheeler and his men were among its official guides. Part of their mission was to inventory the economic potential of the lands they examined, and this they did with insight and thoroughness, to the benefit of ambitious emigrants from the East.

In addition to Blunt, First Lieutenants W. L. Carpenter, Rogers Birnie, Jr., and Philip M. Price also led field parties through northern New Mexico, and their long meanderings afforded them ample opportunity to evaluate the territory's economic resources. Their observations were incorporated into Wheeler's *Annual Report* for 1876, wherein they noted that north-central New Mexico was endowed with "wonderfully good" grazing both winter and summer; that its mountains supported stands of fine spruce and pine timber; that although there were few important mines in the Sangres, "there are indications of silver, gold, and copper throughout the range"; and that the irrigable valleys of the region were fertile and abundantly watered.

Why then, asked Lieutenant Carpenter in a report on his own accomplishments for that year, "with all its natural advantages, its bright skies, pure atmosphere, and healthful climate," why is it "that the emigrant wagon is so seldom seen bringing pioneers to new homes in this desirable region, while other Territories, far less inviting are being populated so rapidly?"

Carpenter answered his rhetorical question with both prejudice and prophecy. He began by restating the widespread Anglo view that New Mexicans were too lazy to make good use of New Mexico:

In no part [of the territory] can anything approaching to eastern thrift and luxury be found. The picture which the writer saw, of a thrifty field of corn in which a herd of cattle was peacefully browsing upon the tender grain while the owners were in-doors taking their noon-day siesta, is but a type of native shiftlessness too prevalent for the common weal.

As a result of New Mexicans' insouciance toward material gain, said Carpenter, the agricultural lands of the territory were by no means fully utilized, and chief among them, the fertile valley of the Rio Grande was capable, by his enthusiastic estimate, "of supporting ten times its present population." Elsewhere in the territory, the possibilities for livestock grazing were virtually limitless:

The writer has visited nearly all the Western States and Territories, and having had a good opportunity for judging of their relative merits for agricultural and grazing purposes, has no hesitation in declaring that as a live-stock country New Mexico is far superior to any other west of the Mississippi. . . . The section drained by the Canadian River and its tributaries, the Mora and Pecos [!], is a fine tract for all kinds of live-stock, and is *par excellence* the future great wool-growing center of the West.

Carpenter pointed out that there was one major obstacle to the realization of New Mexico's considerable potential. It involved "a peculiar state of affairs, affecting the price and ownership of land, which does not exist in any other Territory." Nearly all the land in the territory which might be easily farmed or ranched consisted of Spanish and Mexican land grants whose titles were almost invariably uncertain and required long and costly litigation to protect. Moreover, since the grants were quite large, some of them encompassing hundreds of square miles, individuals of ordinary means could not hope to purchase them. Instead, large, well-capitalized companies were gaining possession of the grants, and their monopolistic control of New Mexico's resources boded ill for the future of the Territory. Lieutenant Carpenter proposed an unconventional solution for this unhealthy state of affairs. "It would be economy for the Government," he wrote,

to buy up all these claims at a good round sum and throw the land open to settlement under the homestead laws. The speedy increase in the population and amount of taxable property, and the general prosperity of the people, would soon more than repay the original outlay, and change this conservative, inert Territory into a thriving rival of her less-favored neighbors.

Carpenter was advocating a kind of paternalistic federalism that had until then hardly been contemplated in the United States. Carpenter believed, and so did George Wheeler, that the resources of the West were too valuable to be left to haphazard private exploitation. Again and again in the *Reports* of the survey, Wheeler, Carpenter, and others argued that the government should play a manager's role in developing the unsettled territories. The United States had used the surveys to build a road west; now the time was fast approaching, said the surveyors, to pave that road, erect signs, and arrange for order and justice in the commerce of its travelers.

The decision makers in Washington, however, ignored the recommendations of Wheeler and Carpenter and would continue to do so for decades more. In 1879 Congress abolished the Wheeler Survey and combined the remaining civilian surveys of John Wesley Powell and F. V. Hayden into the new United States Geological Survey headed by Clarence King. It is reasonable to speculate that had Wheeler been able to continue his work, his view of the federal role in western development might sooner have triumphed and the history of the West been very different. Any change in that direction might have benefited New Mexico, for Lieutenant Carpenter's negative assessment of land tenure in the Territory was, if anything, far too mild. The concentration of New Mexico's best real estate in the hands of a few wealthy men was producing a similar concentration of economic and political power, and New Mexico, instead of achieving greater social and economic harmony as it might have under more democratic conditions, was destined to suffer several decades of extreme corruption and lawlessness.

During those decades, as Carpenter predicted, sheep ranching came to New Mexico on a grand scale, and the brutal annual shearing that characterized that industry describes equally well the effect of U.S. land policies, or lack of them, on native Hispanos. In New Mexico, as the Wheeler Survey folded its tents and withdrew, the fleecing of the villagers began with a vengeance.

12

Fractions of Justice

*I have just been offered $10,000 profit on that
Grant which I bought the other day. I refused. You
see I wanted this pretty bad, and there are now
four other fellows that seem to want it pretty bad
also. I never had anything that I wouldn't sell, so
they may induce me to part with it. This is
confidential.*

—*Frank Bond to a business associate,
February 20, 1903*

SHARP DEALING IN LAND and law was so commonplace in New
Mexico that the first Surveyor General of the Territory began
his tenure with fear lest the shady dealings of the Mexicans
corrupt the moral Americans who came to the new land to
seek their fortunes. He need not have worried. Immigrant
Anglos, quite on their own initiative, soon compiled a record
of unscrupulousness beside which even the worst swindles of
New Mexico's Spanish and Mexican past seemed timid. By
exploiting the discordances between Spanish and American
codes of law, Anglo speculators, often assisted by native New
Mexican *ricos* and politicos, managed to buy up many tens of
thousands of acres of valuable land grants for very little money.
Their success, which stripped most of the Territory's Hispanic
villagers of their patrimony and their chief source of wealth,
yielded a legacy of bitterness that troubles New Mexico even
today.

Land was the hub of New Mexico's economy as it began
finally to roll forward in the second half of the nineteenth

century. Charles Goodnight and Oliver Loving pioneered their famous cattle trail up the Pecos in 1866, and within a few years enough longhorns had been driven from Texas to launch cattle ranching as a major territorial industry. Sheep raising, too, expanded rapidly in the seventies, and logging and mining began to gain momentum after the railroad reached Las Vegas in 1879. None of these industries, however, shaped the future of New Mexico as decisively as the business of land speculation, for land was the treasure on which every other venture depended.

The best real estate in the Territory consisted of the Spanish and Mexican land grants, which featured the choicest grazing, water, and timber in the region and which offered, because of their location within and around the most densely populated areas, the added bonus of being relatively free from Indian attack. Even grants with badly clouded titles were often worth acquiring since the use of a large tract of land through long years of title litigation could yield a fortune to a rancher or lumberman, regardless of whether he ultimately retained possession of it.

Title litigation, because of the inexactness of Spanish and Mexican records and the difficulties of reconciling Spanish law with the tradition of English common law that the Americans brought with them, became a major industry in its own right in New Mexico. According to the Treaty of Guadalupe Hidalgo, by which Mexico ceded its northern territories to the United States, the United States pledged to respect the property rights of all former Mexican citizens living in its new possessions. Accordingly, it was obliged to respect the titles of all valid Spanish and Mexican land grants. Separating valid from invalid grants, however, proved to be a difficult task, and, to make matters worse, Congress failed to provide a procedure for the confirmation of New Mexican grants until 1854, when it appointed William Pelham to be the first Surveyor General of the Territory. Even then the provisions that it enacted were woefully inadequate.

Pelham and his successors never commanded the manpower or the funds to accomplish the enormous task of recording

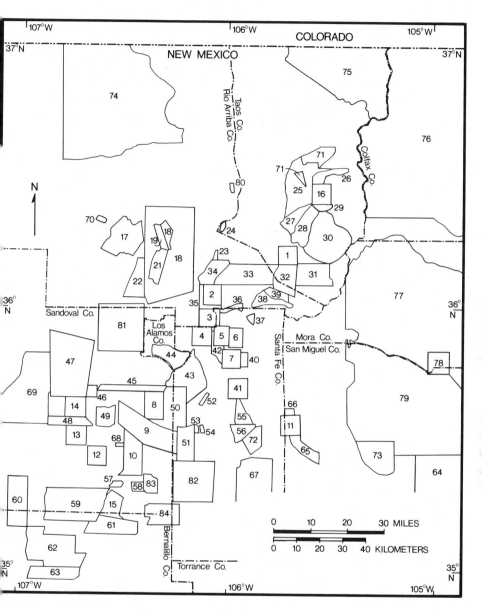

Map 5. Spanish and Mexican land grants in north central
New Mexico. Only grants confirmed by the United States
are shown. See Appendix for data on individual grants.

and evaluating land grant claims. Land grants, in fact, were not thought to warrant much of their attention. For as long as there was a Surveyor General in New Mexico, his main duty was to survey and administer the territory's public domain (its unsettled, unclaimed lands). Dealing with the endless title problems of the land grants, where the majority of the people of New Mexico lived, was considered a secondary occupation to which he might attend during lulls in his other work. As a result, between 1854 and 1891, when Congress finally created the Court of Private Land Claims in order to resolve the land grant problem, only 22 of the 212 claims recorded by the Surveyor General had been patented. This meant that for thirty-seven years, more than thirty-five million of New Mexico's best acres were in a legal limbo.

Through the long periods of delay, the unsophisticated Hispanic farmers who lived on the grants were easy prey for both sharp Anglo lawyers and unscrupulous New Mexican *dons.* The difficulties they faced in protecting their interests were immense. They neither understood the American legal system nor spoke its language, nor did they have friends among the businessmen and politicians who made the system work. There was no shortage of lawyers to promise them guidance in the presentation of their land grant claims to the Surveyor General, but in the process of unraveling the difficult legal tangles, the lawyers inevitably gained possession of a major portion of the villagers' land. Some lawyers resorted to such blatant chicanery as inducing a client to sign a document, described as having no particular importance, which later turned out to be a quitclaim deed to the client's land. More often, however, a lawyer acquired his land by legal means. He had to be paid for his services, and since the subsistence farmers of the grants generally had no cash, the lawyer was paid with a one-third or one-quarter interest in the grant.

Cash, or lack of it, presented a number of other problems. The property taxes levied by New Mexico's new county governments had to be paid in cash, and from time to time so did the costs of surveys and other official services. Moreover, cash was the key to enjoying the manufactured goods that rolled out to New Mexico from the factories of the East. The common man suddenly found that he needed and wanted cash in more ways than he had ever dreamed possible, and besides his labor, which brought him little, he had only one way of acquiring it—by selling his land.

In the drama of the land grants the actors opposite the Hispanic farmers were wealthy lawyers, politicians, and entrepreneurs who had access to giant amounts of capital and who fully understood that the expanding web of railroads in the territory would cause land values to increase and their investments to boom. These speculators, however, are not alone to blame for the hardship and injustice that land grant chicanery produced. Probably the greatest culprit was the United States Congress, which up to 1891 failed to establish adequate legal procedures for adjudicating land grant claims and neglected to furnish sufficient instructions for easing the transition from Spanish to American law. This last omission produced particular pain with respect to one of the most favored devices for separating Hispanos from their land: the suit for partition, which was used to arrange the sale of major portions of most of the land grants in the southern Sangres, including the Las Trampas Land Grant, one of the oldest and best documented grants of all.

During the relatively peaceful decades of the early 1800s, settlers from the tiny village of Las Trampas moved out from their fortified plaza to occupy all the available arable land within the grant and to establish the new communities of Ojo Sarco, Chamisal, El Llano, Cañada los Alamos, Ojito, and eventually El Valle, a village that became particularly entangled in the legal affairs of the grant. Because the likelihood of Indian attack was no longer very great, the houses of these new settlements were not grouped close together but were scattered along the edges of the irrigated fields so that each family might better guard its crops from livestock and theft.

In every village on the Las Trampas Grant, as on every community land grant, farmland was apportioned among the original settlers in roughly equal amounts, each head of household

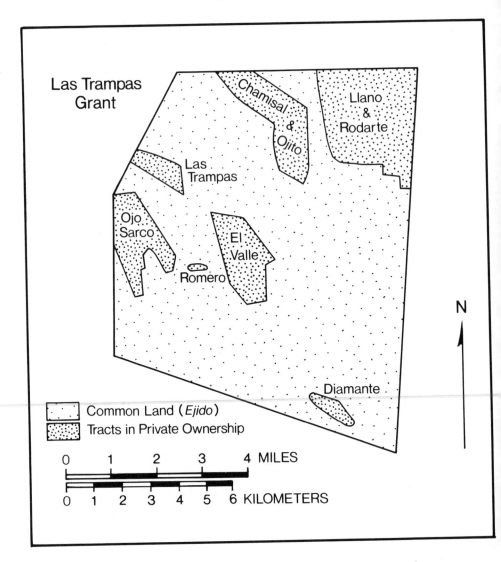

Map 6. *Ejido and village tracts of Las Trampas Land Grant.*

receiving a certain amount of river frontage. From the river his property extended in a strip of even width across the irrigable bottomland and up through the dry hills, where the houses and barns were built, and on still farther to the tops of the ridges that defined the immediate valley enclosure of the village. These strips of land, which became progressively narrower as they were divided among successive generations, were acknowledged to be private property that could be freely bought and sold.

Beyond the ridges that enclosed the privately owned land of the villages lay the land grant commons, the ejido. This was the land that speculators eventually coveted and that the American courts permitted them to buy and sell. Under Spanish and Mexican law, however, the ejido had no marketable title. It belonged to the members of the grant community as a whole, and although they could assign parts of it to new settlers for their houses and corn fields, the vast forests and rangelands of the ejido could not legally be sold or otherwise alienated from the community. A major cause for the injustices associated with land grant speculation in New Mexico was the failure of the United States to recognize its obligation to uphold this provision of Spanish and Mexican law.

The lands of the ejido were open to all the people of all the villages on the grant to use as they saw fit. On the Las Trampas Grant, these lands ranged from the dense spruce and fir forests on the slopes of 12,200-foot Trampas Peak, down through mid-altitude stands of ponderosa pine, and finally to the rough, semi-arid hills of the piñon-juniper woodland. The ruggedness and limited productivity of the mountain country created an ecological niche that the small villages of a community land grant were well equipped to fill. Arable lands were few and widely scattered. The best firewood lay at one elevation, the best house timbers at another, and the location of good grazing varied with altitude and with season. Overall, there was not enough of any single resource to support a group for long. In order to eke out a living the people of the Las Trampas Grant had to make full use of every available resource, and they had to cooperate with each other to do it. Only by sharing their goods and their labor could so isolated a people, possessing so few tools, manage to survive in as unforgiving an environment as the southern Sangres.

In the matter of being recognized as the legal owners of their land, the people of the Las Trampas Grant got off to a good start. Surveyor General Pelham reviewed their claim in 1859 and soon afterward recommended that Congress confirm it, which it did on June 6, 1860. The Las Trampas Grant seemed to be navigating the correct legal channels with admirable speed and luck, but still the problem of an accurate metes-and-bounds survey remained, and the intervention of the Civil War and Reconstruction delayed all efforts to obtain one for fifteen years.

As was commonly the case in New Mexico, the boundaries of the grant had originally been set forth in only the vaguest terms. Locating the eastern boundary, which was cryptically recorded as "the little canyon of the Rito San Leonardo at the mountains," proved especially troublesome. In fact, the General Land Office in Washington rejected as inaccurate four successive surveys of the grant before it finally accepted the work of Clayton Coleman in 1893. From the rough rectangle two and a half leagues on a side evoked by the instructions of Governor Vélez Cachupín in 1751, the grant had metamorphosed into an irregularly shaped pentagon in which the American government counted 28,131.66 acres. It is doubtful, however, that this transformation mattered very much to the fifteen hundred or so inhabitants of the grant. For them life was a slowgoing pastoral experience, and the business of measuring the land must have seemed a strange and amusing form of lunacy.

Unfortunately one of the many inheritors of the grant was less than peacefully reconciled to the condition of his estate. He was David Martínez, Jr., who lived not on the grant but near it in the town of Velarde, and he was chronically in debt. In 1891 he borrowed $1,000 from the First National Bank of Santa Fe and promptly defaulted three months later when the loan came due. By September of 1892 he had paid back only $30.46 of the principal, and although the bank had made the loan without collateral, he knew that he was in trouble and that his trouble was deepening at an annual rate of 12 percent.

In November 1891 he was already exploring the possibility of suing for partition of the Las Trampas Grant, in which he, of all the descendants of the original 12 grantees, held the largest single interest. The suit for partition was a peculiarly Anglo contribution to land grant litigation. The idea behind it was that the commons of a land grant was simply the aggregate of a large number of individual possessions. Therefore, if only one out of 10,000 holders of a grant desired, he could sue the other owners for his share of the commons, and if the physical separation of his fractional tract from all the others was impractical (and it nearly always was), then *the entire grant might be sold by order of the court* in order to divide among the former owners of the grant the cash equivalents of their shares. This warped interpretation of the essential character of a community land grant would have been unthinkable under Spanish law, but in New Mexico legislation to inhibit it was not enacted until 1913. By then most of the territory's community land grants had already been alienated from their original Hispanic owners through partitioning or some other means.

Because of various bureaucratic delays within the General Land Office, Martínez had to wait eight years before initiating the legal process that he hoped would convert his patrimony into dollars. On October 18, 1900, in association with four other descendants of the original grantees, he filed suit for partition of the Las Trampas Grant, "and if partition cannot be made . . . , then for a sale of said premises and for a division of the proceeds thereof." His attorney in the case was Alonzo McMillen, a young lawyer from Albuquerque, whose handling of the case over the next several years would evidence more avarice than ability.

At first the suit moved swiftly. The defendants, who included all the descendants of the original grantees except Martínez and his cohorts, were summoned to court by means of newspaper advertisements published in English. When none of them appeared (because of those who could read, few read

English, and of those none read the legal advertisements in the Taos newspaper), the court judged that by their absence they had consented to the appointment of a referee to investigate the question of partitioning the grant.

The referee appointed was Ernest A. Johnson, who in April 1901 opened hearings in the mountain villages located on the grant. Essentially, Johnson's job was to determine the identity of the grant's owners and the size of the share of each in the commons. His assistant throughout the hearings was Amado Chaves, who was a silent partner with McMillen in the latter's burgeoning interest in the grant. Johnson was also responsible for determining how much of the grant was commons and how much, by reason of occupation and the making of improvements, had become the private property of the various settlers living on the grant. Incredibly, Johnson recorded that a mere 650 acres of the grant belonged to the latter category. This preposterous figure, which was not one-tenth as great as it should have been, suggested that fifteen hundred people found space in one square mile of ungenerous mountain land to cultivate their hay, grain, and vegetables, winter their livestock, and build their houses and barns as well. The paucity of land credited to the settlers would bring ample grief to future owners of the grant, but none of it, unfortunately, would be visited on McMillen and his deserving crew.

Referee Johnson compiled genealogies for every tenant of the grant and computed the fractional share of each in the commons. He was guided in his efforts by the dubious principle that the more directly an individual was descended from the original grantees, the more land he was entitled to. Johnson discovered from his forest of family trees that David Martínez was the owner of 1187/6480 of the commons (strangely he did not use percentages, but this unwieldly number works out to 18.3%) and that the shares of other owners ranged from as little as 1/14,000 to as much as 2,966,813,402/27,861,926,400 (10.6%)—this last mathematical nightmare representing the interest of Alonzo B. McMillen, who having gained a quitclaim from Martínez for 1/4 of his share, was fast acquiring deeds to 1/3 or 1/4 interests in the shares of his other "clients."

The absurd list of names and fractions was carried back to Santa Fe, where the presiding judge, Daniel H. McMillan, reviewed them and speedily appointed a board of three commissioners to effect a physical partition of the commons of the Las Trampas Grant. The commissioners, Ireneo Chaves, Elias Brevoort, and Henry W. Easton, were selected because none of them was connected with any party to the suit "either by consanguinity or affinity." Apparently Judge McMillan was not impressed by the fact that Ireneo was Amado Chaves's brother. Such a fine point, however, hardly mattered; the conclusion of the commissioners was foregone before they ever swore their oath of office.

Photo 41. Las Trampas in 1915, as it looked in the days of the land grant struggle. The magnificent church which Juan de Arguello and his fellow pioneers commenced building in 1766stands at center. One of the first barbed-wire fences in the village is barely distinguishable in the middle ground on the right. Photo by Jesse L. Nusbaum. Courtesy Museum of New Mexico.

It came as no surprise that on September 20, 1901, the three commissioners returned to court and confessed that "owing to the character of said lands and to the large number of owners (there being nearly 300) and to the fact that many of the interests are very small," a physical partition of the grant was utterly impossible. Judge John McFie, who had replaced McMillan, then decreed that since partition could not be made, the entire grant, excepting an aggregate of 650 acres for settlements, "be sold at public auction at the front door of the Court House in the City of Santa Fe, New Mexico, to the highest and best bidder for cash."

McFie appointed Ernest Johnson as Special Master to conduct the sale, and accordingly advertisements ran for four consecutive weeks in the Taos *Cresset* during January of 1902. As the day of the sale drew near, the scent of riches stirred new activity among the Santa Fe business community. Roman L. Baca and a number of associates, who included some of Santa Fe's most prominent merchants, were creditors of Amado Chaves in the amount of $13,690.66. Twenty-four hours before the opening of the sale, the Baca group filed suit against Chaves, McMillen, and Johnson in order to attach a lien on Chaves's interest in the grant. Although the sale went ahead as planned, the Baca suit would soon have consequences for everyone involved.

To the distinct pleasure of some of the participants, the sale was not well attended. The only bid made was one of $5,000 offered in the name of H. F. Raynolds by Raynolds's law partner, the ubiquitous Alonzo McMillen. Special Master Johnson closed the sale, drew up the appropriate papers, and the affair seemed to be racing toward its conclusion. Two days later Judge McFie convened a post-mortem hearing at which he confirmed the sale, ordered that the cost of the Coleman survey and the fees of the commissioners, referee, and special master be paid, and generously set McMillen's fee at one-quarter of the interests of all the owners of the grant with whom McMillen had until then not managed to arrange a contract. In other words, several hundred people who had never met nor hired McMillen were suddenly obliged to pay to him one-quarter of the proceeds of something which they had never intended to sell. Moreover, although in theory they were paying McMillen for his service to them, his only service had been to himself in arranging the sale of the grant at far less than its fair market value to a group of investors of whom he himself was a member.

It is alarming to note that up to this time, not one of the defendants in the case (that is, the owners of the grant, exclusive of Martínez and his four friends) had been represented in court. No motions had been filed and no documents signed in their behalf; indeed, they had no attorney. And yet their collective patrimony had been sold from under them at the paltry price of eighteen cents an acre. Later events show, in fact, that the residents of the grant did not even know that these machinations were going on. They were being roughly sheared, and did not feel a thing.

Incredibly, the day was at least temporarily saved by the creditors of Amado Chaves, who moved to have the sale declared null and void because they feared they might not get their money if the grant sold at so low a price. Their attorney, Alois B. Renehan, catalogued a dozen reasons why the sale should not be allowed to stand, but chief among them was his contention that before McMillen and Raynolds had purchased the grant, they had already contracted to resell it for $1.50 an acre. Renehan further alleged that Johnson had been bribed to comply with the scheme and that therefore all three were guilty of a conspiracy to defraud the owners of the grant. Judge McFie must have found some substance in Renehan's objections, for on May 2, 1902, he canceled the sale to Raynolds and ordered that a new one be held (with Johnson still as Special Master).

David Martínez must have been relieved. If the sale had gone through, nearly a quarter of the auction's proceeds of $5,000 would have been paid out for survey and court costs; another $1,000 would have subtracted for back taxes, and out of the remainder Martínez would have been lucky to net $380, which was not nearly enough to rescue him from indebtedness. And his situation was growing more and more desperate. As 1902 progressed, Rio Arriba County served notice that it would seize a one-hundred-acre tract belonging to Martínez (probably his home in Velarde) for nonpayment of taxes. Worse, the First National Bank of Santa Fe, still pursuing payment of the 1891 promissory note, had filed to obtain a lien on Martínez's interest in the grant. Martínez gamely fought back through a succession of seven separate legal actions (while his legal costs soared), but all his efforts were in vain. In February 1903, Judge McFie named the bank a successor to Martínez's interest in the grant, up to the amount of $1810.46. Suddenly Martínez, who had initiated the long and complex process of partition, found himself one of its less essential actors. His position became all the more ironic when the patent for the grant, bearing the signature of President Theodore Roosevelt, was finally delivered only a few days prior to the bank's victory.

With the problem of Martínez's indebtedness out of the way, the path was clear for a second auction of the grant. The event was set for 10:00 A.M., February 7, 1903, at the front door of the Santa Fe courthouse. On that morning Martínez came early and pleaded with Johnson to stop the sale. He said that he would not let go of his share of the grant for anything less than a dollar an acre. Claiming that he knew of buyers who would meet that price, he insisted that the sale be postponed a few days more. Johnson, however, proceeded with the auction.

The highest bidder that day was Frank Bond, an Española sheep dealer and merchant, who purchased the grant for $17,012, or about 60¢ an acre. Out of that amount at least $2,500 went for taxes and official costs, and $4,280 was consumed by McMillen's fee. After First National collected its money, David Martínez stood to receive about $200, with which he might have paid his taxes and lawyers and barely had enough left over to drown his sorrows at a cheap cantina. As for the other three hundred or so inheritors of the grant, their shares should have been distributed from the remaining $8,200 by their un-hired attorney Alonzo McMillen. The average share they received, if McMillen did not pocket the money, was about $25, which was not enough to buy a set of harness at Frank Bond's store.

The completed sale marked the beginning of a fresh chapter in the history of the Las Trampas Grant, and several years passed before the subsistence farmers of the mountain villages began to realize what had been done to them. The new owner of the grant, Frank Bond, was one of northern New Mexico's leading entrepreneurs. In 1883, as a recent immigrant from Canada, he had settled in Española, New Mexico, which a few years earlier had been joined to Colorado and the rest of the world by the tracks of the Denver and Rio Grande Western Railroad. Along with his brother George, Frank Bond foresaw a lucrative connection between rails and sheep, and within a short time the brothers' retailing and wool-growing operation had spread over northern New Mexico and well into Colorado. As their incomes swelled, they speculated increasingly in land grants and logging, and the Las Trampas Grant, which Frank rated as "undoubtedly the best timbered grant in this section of the country," was a prized addition to their holdings. Still, they had no wish to keep it merely for the sake of ownership.

A few weeks after the courthouse auction, the Bonds tentatively offered to sell the grant for $1.50 an acre, but almost immediately Frank had second thoughts and wrote to his brother describing a new plan for capitalizing on the grant:

My idea is that we will not offer it to anybody until we offer it to the U.S. . . . If you will look at a map of N.M. which shows the grants, you will note that the Trampas Grant cuts into the Pecos Forest Reserve. I think on account of its location, being at the head waters of several streams, and also being joined onto the Reserve, that we could trade it to them for Scrip, provided that we can work with the right kind of people. Middlecoff states that it would take money to make the sale. He mentioned that these officials would have to be approached, in such a manner that they would know it would be to their personal advantage to make the deal.

Scrip was widely used by the government in place of cash in

land transactions. Scrip issued in exchange for land in New Mexico might be redeemed for acreage elsewhere in the public domain—often at a sizable profit, as Bond learned from Senator T. B. Catron's son Charles:

Young Catron informed me on the train about three weeks ago that his father sold the U.S. part of the Gabaldon Grant (which is on the head waters of the Santa Fe River near Santa Fe). This also cut into the Pecos Reserve. They only traded them 2,000 acres. He took this scrip and went up into Washington Territory and located some of those fine timber lands there. He claims that he could sell this timberland in Washington Ter. at $40.00 per acre. Young Catron is very windy and probably did not tell all the truth.

Frank Bond, the merchant, however, unlike the lawyer/politicians of the Catron family, was somewhat at a loss as to whom he might employ to exert the necessary influence:

It might be that Catron would put us on to how to make this trade with the government so that it would not be very expensive. This is well worth looking into. Whom do you think we could approach? There is Governor Otero and Rody [sic], both of whom are after all there is in it, but not very reliable people to do business with, but we would not give up anything until we had our Scrip, or until we sold it.

Bond wrote to the commissioner of the General Land Office in Washington in April of 1903 suggesting the trade, but his efforts came to nothing, possibly because he never found "the right kind of people" to work with.

Finally, in 1907, a group of four Albuquerque businessmen became sufficiently attracted to the grant to incorporate themselves as the Las Trampas Lumber Company and to make a down payment of $16,000 (the final purchase price remains unknown). To their great dismay, however, they found almost immediately that the title to their new possession was not secure. Inhabitants of the grant had at last learned, probably through employees of the lumber company, that they no longer were the owners of the mountains and hills where they grazed their flocks. They further learned that some of them no longer even owned the houses they lived in. The situation was particularly serious in the village of El Valle, where Ernest Johnson (who now was working for Frank Bond as a traveling field representative and troubleshooter) had excepted only fifty acres from the surrounding commons. In actual fact, the privately owned land of the village sprawled over fifteen hundred acres. Almost all of it had been sold!

Juan B. Ortega of El Valle resolved to resist the new state of affairs. Besides being a farmer and rancher, Ortega also occasionally functioned as a "lawyer" in minor cases brought before local justices of the peace, and doubtless his reputation as a man familiar with legal issues stood him in good stead as he recruited other elders of the village to join him in opposition to the Las Trampas Lumber Company. At dawn one morning in 1908 he and his group climbed together into a single wagon and embarked on the fifty-mile journey to Santa Fe. Once there, they set about to find a lawyer to represent them, and the man they finally decided to hire was none other than Charles C. Catron, the "windy" son of T. B. Catron, the greatest land grant speculator of them all.

Probably it was Catron who informed O. N. Marron, the attorney for the Lumber Company (and also one of its partners), that the villagers were determined to resist the company's claim to ownership of portions of the grant. Marron quickly responded by firing off a pair of suits. One was an action to quiet title directed against Ortega, his group, and anyone else who might contest the company's claim to the grant. (A quiet-title suit is a legal action designed to settle definitively the ownership of a piece of real estate.) Marron's other suit called upon Frank Bond to fulfill his duties as warrantor of the company's deed to the grant. Frank Bond in turn reached for his lawyer, who turned out to be a man well acquainted with the background of the case, Alois B. Renehan.

Not much happened in either of the suits for quite a while, but eventually, in 1911, the villagers, through Catron, formally stated their case. Later events would show that most of the villagers who were aware of the suit thought that it involved a defense of their actual ownership of the commons, before partition. Possibly not even Ortega fully understood that their

There are many other qualities present—vigorous thought and cultivated style, but fearless recognition of facts and intense earnestness will still remain his great characteristic

He is irresistible before juries

HON. A. B. RENEHAN.
SANTA FE

Photo 42. Alois B. Renehan. Cartoon from the Las Vegas Optic, no date. Courtesy Museum of New Mexico.

claim was much more limited and that all that they, through Catron, were alleging was that private lands had been swept up in the process of partition and unlawfully taken from them. As anyone could see who visited the grant, which C. C. Catron did, their cause was just. Nevertheless, the wheels of the law turned slowly.

Part of the delay was caused by wrangling between the company and Bond. The two did not reach an agreement on how to handle their mutual problem until May 1913, when Bond agreed to pay $57,650 for half the stock of the company and assumed full responsibility for solving the muddle of the title. Bond and Renehan promptly hired a surveyor to map the exact dimensions of the cultivated and inhabited tracts on the grant, and Renehan and Catron worked out an agreement. What they decided was this: the descendants of the original grantees— every one of them—"for $1.00 in hand and other valuable consideration" would execute quitclaims to their interests in the commons, and the company, for its part, would abandon its claims to the private tracts. Further, the company would respect the rights-of-way of irrigation ditches that flowed through its land, and would permit the villagers to graze their livestock on the former commons and to harvest its unmerchantable timber for fuelwood and domestic building needs.

During the summer of 1913 the new survey was completed, and it showed that 6,950 acres, and not the laughably inadequate 650, should have been excluded from partition. That same summer Catron and Renehan went up to the grant to explain the agreement and secure the quitclaims. Their job was a difficult one, as it required locating and talking with several hundred individuals who had every reason to be reticent and suspicious. After only a few days Renehan tired of roughing it in the mountains and left Catron to do the job alone. Catron in turn relied on Juan B. Ortega, at least in El Valle, to guide him on his rounds and to help him persuade

Photo 43. El Valle. The fields shown here constitute about a third of the cleared land in the village. Author photo.

the head of every household to sign the obligatory quitclaim.

The suit dragged on for half a year more as the names of residents and claimants continued to trickle in. Finally, in the spring of 1914, with Catron's fee of $5,500 having been paid by the Las Trampas Lumber Company, the long legal battle came to a definitive end. Bond immediately put the grant up for sale for $160,000 (nearly a tenfold increase over the purchase price of 1903) and began to advertise that it possessed eighty-five million board feet of saw timber and one million ties. In writing to one of his partners Bond gave some indication of his feelings for the Hispanic villagers and the antiquity of their settlements when he mentioned that a prospective buyer might want to see "a map of the Trampas Grant showing what lands have been surrendered to the squatters."

Although the grant attracted a few inquiries, it did not sell. Reluctantly, the Las Trampas Lumber Company went into the lumber business. A sawmill was built in El Valle and a shipping station constructed next to the railroad at Velarde, but the manufacture of ties, poles, pilings, and boards never prospered. In 1926 the Las Trampas Lumber Company was obliged to declare bankruptcy. Later that year its receiver accepted a three-party swap in which George E. Breece Lumber Company bought the grant for $62,320 and straight away traded it to the U.S. Forest Service for $75,000 worth of standing timber in the Zuni Mountains west of Grants, N.M. In this way, with no expenditure of government money, the commons of the Las Trampas Grant finally became the property of its present manager, the Carson National Forest.

The partition and sale of the Las Trampas Land Grant had proved to be a bitterly ironic affair. David Martínez, the eager bungler who initiated the whole sorry business, emerged from it nearly as debt-ridden as he had started out. His personal ambition had cost his countrymen their greatest single source of wealth. A. B. McMillen and Amado Chaves, on the other hand, substantially enriched themselves by their shearing of the Tramperos and went on to amass a still greater fortune by relieving the unlucky residents of the Cañon de San Diego

Grant of their commons. More than Frank Bond or any other player in the drama, they managed to escape from the Las Trampas Grant with their profits intact, and they, least of all, deserved to do so.

The cruelest irony, however, was visited upon Juan Bautista Ortega, who had struggled to win back the farms and homes that had been auctioned along with the commons. For his efforts he became known, not as the man who had rescued the village from disaster, but as the *traitor who sold away the grant.* Apparently almost no one in the village realized that the commons had been lost in 1903. According to oral tradition, the Las Trampas Grant was not sold until ten years later when Ortega and Charles Catron went from house to house dispensing dollar bills for quitclaims. Ortega, some said, accepted the full amount of the purchase price from Frank Bond (who was well known to the villagers as a merchant) and then distributed only a fraction of it, dollar by dollar, to the various heads of families. Supposedly he kept the rest of the money for himself. It seems that virtually no one, not even the other men who traveled with Ortega to consult with Catron in Santa Fe, understood fully that their homes and farms, as well as the commons, had been unfairly conveyed to Bond in 1903.

The confusion about the grant was not the only problem besetting Juan Ortega and his neighbors. During the early years of the century more and more men regularly left the villages of northern New Mexico to earn cash wages as sheepherders, section hands, and miners. The money and ideas they brought back from their employment caused a permanent transformation of their home communities. Although too isolated to receive many of the benefits of American society, the villages of the Sangre de Cristo were thoroughly exposed to the pressures of the cash economy and the individualistic, get-ahead ethic that drove it.

The new circumstances required the villagers to make difficult, wrenching adjustments, and, having lost their land grants, they had lost the one useful economic resource that might have afforded them protection and security through the years of rapid change. They were now a landless people living in a great and spacious landscape. They might work as loggers or herders, but they did not own the forests and rangelands that surrounded them. Villagers on the east side of the mountains rose up to protest the new state of affairs. Calling themselves Las Gorras Blancas (White Caps), bands of Hispanic night riders cut fences and burned barns. But their efforts were futile; in a short time, they, too, adjusted to their dispossession.

The experience of the Pueblos, meanwhile, was the opposite of their Hispanic neighbors'. Beginning in 1913, the federal government prohibited Pueblos from selling any of their tribal lands and even undertook a vigorous program of restoring to the Pueblos lands that had been alienated prior to 1913. The Hispanos of northern New Mexico had every right to view these initiatives with some amazement. They too had become citizens of the United States unwillingly. And they too had been naive and vulnerable to the ways of Anglo law. But unlike the Pueblos, they received not the least protection. In the day of McMillen and Bond, as in the day of the Comanche, they were entirely on their own.

13

Manitos

Mountain villages:
under the piled-up snow
the sound of water.

THE DEEP LINES that furrow Jacobo Romero's face read like a map of the rough country where he has lived his eighty-plus years. He is old enough to remember the day when his father, dressed in the suit he reserved for weddings and funerals, drove off with Juan Ortega and the others to hire a lawyer in Santa Fe. His wife, Eloisa, is younger than he and, unlike him, has no recollection of the commotion over the land grant. Her memories of that time revolve instead around the exhilarating terror of a solar eclipse, and the way the rough wooden floor of the Trampas Church used to hurt her knees during mass. Together they look back over the intervening years like sailors who have landed in a place so different from where they started that comparing one to the other makes both sound implausibly exotic.

When they were young, the Hispanic villages of the Sangres were a separate world where it was extremely rare for anyone to see a gringo, let alone talk to one. Now the old couple have several for sons-in-law. In the days of their youth, Spanish was the only language they heard spoken, but now a majority of their grandchildren and great-grandchildren speak mainly English, and many speak it with the same urbane inflections one hears on television. Other old people in the mountain villages have found these changes demoralizing and painful, but not these two. They travel from one end of the country to the other to visit their children, and they find the present to be as satisfactory as the past. Still, they say, there is no denying that much of value has been lost. They consider themselves fortunate to have known the old days when the mountain villages were self-contained and nearly self-sufficient, and when life, moving at a horse-drawn pace, had all the stability—and hardship—of the mountains themselves.

One of the most striking aspects of life in those days was the absence of fences in the villages to separate individual plots of land. Instead of barbed wire, children were employed to keep the livestock where they were supposed to be and to fetch them when they wandered too far. There was only one village fence, made of rails or of upright stakes bound closely together, and it encircled all the cultivated land with no distinction for who owned what. During the summer all the village livestock—sheep, goats, cattle, and horses—were kept outside the fence to graze freely in the mountains and hills under the guard of small boys and their nervous herd dogs. In autumn after the crops had been harvested, the animals were driven back inside the village enclosure to graze wherever their grass-snuffling noses led them. They ranged from one end of the village to the other chomping on corn stalks, spilled grain, and whatever grass had grown after haying, and no one attempted to restrict them to the fields of their owners.

To be sure, there were injustices in the system. A man who farmed a large acreage and owned only a milk cow and a team of horses had no way of reserving the abundant stubble of his wheat fields for his animals' exclusive use. By the same token, his neighbor who farmed little yet owned a large herd of sheep might reap the benefit of far more feed than his own land provided. But these imbalances—and they were recognized as such—helped to hold the community together. They meant that a family owning no land besides that on which its house stood might still keep a few cows and horses and thereby share in the prosperity, however limited, of the village. Today that kind of economic communality is all but unthinkable. As one oldtimer from Mora put it, "People now are more selfish than they used to be. Today every monkey has his swing, and he just stays on that." It may be just as well. The old ways, while generous to humans, were exceedingly hard on the land. Neither the ejido nor the cultivated tracts received any rest except when they were buried under snow, and ultimately the villagers had to pay the costs of overuse.

The mountain villages of the old days were hardly idyllic. Frequent dry spells during the growing season produced heated arguments about the management of the irrigation ditches, many of them fanned by bitter, long-standing feuds. As the story of Juan B. Ortega illustrates, misunderstanding and narrow-mindedness could sour a man's life, and the habit of thinking about the outside world could as easily bring him to grief as save him from it. Yet for all their faults, the villages possessed an integrity and unity that is altogether lacking in most modern communities. Equality of condition and station ensured that villagers were only minimally concerned with private gain, and their ethics reflected that attitude. The most important civic virtue for a man to have was *verguenza*, a self-effacing probity that restrained him from advancing himself at the expense of others. (In reference to women *verguenza* had a different meaning, connoting sexual morality and modesty.) A man who possessed *verguenza* would never enter the predominantly Anglo professions of banking, law, or trade, which required that one profit from others' loss, but he could be a farmer, herder, or wage earner, which of course most villagers were. Such a man would count tradition and rootedness as life's primary treasures. Compared to his sense of fellowship with his neighbors, getting ahead was of negligible value. And besides, there was usually no place to get to.

The villagers lived in isolation from the outside world and drew on local resources for the things they required. On foundations of mud and river cobbles, they laid up the walls of their houses with sun-dried adobe bricks and mud mortar. Their roof beams and lintels they cut from the surrounding forest; their ceilings were herringbone arrangements of sapling poles overlaid with brush, leaves, and ultimately a thick layer of dirt; and for plaster inside and out they used more of their one inexhaustible resource—mud, sometimes dressing up the final coat with the addition of sparkling micaceous sand. From these humble materials they obtained a house that, although prone to leak and always hard to clean, was warm in winter and cool in summer. Moreover, it had a gently organic, sculptural quality: every window and door opening had smooth, hand-rounded edges; walls were not quite plane, and their modest undulations gave them a fluid grace in spite of their great thickness and bulk. The house belonged to the landscape as truly as the tan dirt of the yard before it or the weeds that grew in its roof. When it was abandoned and no one any longer repaired its walls with fresh plaster or shoveled its roof free of snow, the entire structure returned to the soil as completely as it had once come from it.

Until the 1880s, the villagers turned their fields with wooden plows made from sharpened forks of Gambel oak. By making available metal plows from the Midwest, the railroads set in motion an agricultural revolution of lasting consequences, one that the rest of the country had experienced two or three generations earlier. Not only were the new farm implements more durable and efficient, they also had a much lighter draft, which made it possible for a farmer to replace his oxen with horses. The faster-gaited horses, in turn, were also better adapted to the use of other kinds of labor-saving agricultural machinery including mowers and rakes. As quickly as they could accumulate the necessary cash, village farmers replaced their old equipment with the same factory-made harness and farm gear that was used in the East, but there was never enough cash for everything and the conversion from subsistence to industrial farm technologies took decades to accomplish.

Even after the turn of the century the only kitchen staples not locally produced were coffee, sugar, baking powder, and matches. All the rest of their food the villagers either grew or bartered for. Farmers from low and relatively warm altitudes traded their surplus fruits and vegetables—especially chile—

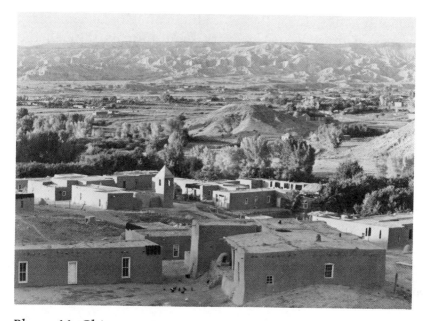

Photo 44. Chimayo, N.M., c. 1912. Courtesy Museum of New Mexico.

for grains grown higher up in the cool mountain valleys. Wheat was the most important crop in the villages of the Sangre de Cristo, and after harvest the sheaves were spread on clean hard ground and threshed by the hooves of sheep and goats driven over it. Log gristmills, perched over mountain streams, ground the grain into flour and loudly groaned and rattled day and night through the busy harvest season.

Regardless of whether it produced mainly fruit or flour, each community was obliged by population pressure to utilize every scrap of available land. Where today village fields grow only hay, they once were thick with corn, wheat, barley, oats, chickpeas, pinto beans, and potatoes. Only fields that were too wet or too rough to plow were given over to pasturage. No one could afford to do otherwise, and in their stores of bran, oats, and corn stalks, the villagers laid up enough supplementary fodder—in most years—to carry their animals through the winter.

The old people of the villages remember the days of self-sufficiency with pride. At least in memory, the poverty they experienced was salubrious. Every man was his own master and the equal of his neighbors. Although manufactured goods were scarce, no one went hungry. There was always plenty of lamb to eat, as well as mutton and goat and sometimes venison from a deer that the dogs chased and exhausted in a winter snowdrift. There were wild turkeys, many more than today, and chicken, pork, beef, trout from the river, grouse, and rabbit. People were poor in material comforts and their children slept crowded together, several to the bed, but they were rich

Photo 45. M. R. Martínez and family, Chimayo, c. 1895. Ristras of red chile and drying lamb or kid hang from the portal. Courtesy Museum of New Mexico.

in what they valued most: time, family, and the freedom of the land.

There was extravagance in the old customs that the villagers' ancestors had brought up the trail from Mexico. Several times a year each community held protracted celebrations of feasts and saint's days. One of the most important was the feast of the patron saint of the village, when the santo, the statue representing him, was taken from the church and carried at the head of a procession to every corner of the village to bless the fields, the irrigation ditches, and the ditch gates. Although their feeling for the mountains and mesas that enclosed them may not have been as profoundly religious as that of their Pueblo neighbors, the Hispanic villagers showed in this custom a deep reverence for the land of the village itself.

There were also long, morbid rituals of death and joyful ones of marriage and christening. Courtship, most of all, was an intricate matter. The betrothal of Jacobo and Eloisa, like most, depended on a lengthy series of meetings between the two families. Haste would have been an admission of poor breeding. Jacobo's father posed the question of his son's marriage to Eloisa only after a suitable number of visits had taken place and after all the prospective bride's many relatives had had an opportunity to meet the parents of the suitor. Eloisa's family received the request with the utmost gravity and formally consented to consider it. Then Jacobo and his family waited. If a messenger from Eloisa's family came to their house within four days, then the news was bad—he had been rejected. But if the messenger did not come until the fifth day, or even later, then the answer was yes, and he would be married. There was no way to judge what the verdict would be as the four interminable days crept by. But they did pass, and no messenger came, and so the wedding, with all its attendant dressmaking, baking, butchering, and nervous parleying with the minister, was on.

The wedding picture (which was taken some time after the wedding when a relative borrowed a camera and persuaded the couple to dress in their finery again) shows a serious young man with his hair cut so short it bristled, wearing a simple dark suit and standing beside a very thin, pretty, dark-eyed teenager in lacy white. She seems unsure what to do with her feet. Following the wedding ceremony, Eloisa's family gave a jubilant feast that was attended by virtually all the residents of Las Trampas and El Valle, as well as by many more friends and relatives from other nearby villages. As tradition demanded, the couple then went to live the first eight days of their married life at the home of Eloisa's parents, and only after surviving that period of probation and scrutiny was Jacobo allowed to take his bride to his own family's home, where an even greater round of celebration awaited them.

The rich customs of the mountain world depended on the luxury of abundant time. Especially when winter snow and spring mud rendered much of the mountain country inaccessible and the somnolence of the land enforced long periods of inactivity, northern New Mexico deserved its reputation as the "land of mañana." But that was only part of the year. At other times there was so much to be done—plowing, irrigating, harvesting, herding, hauling firewood, and repairing the irrigation ditches—that there was not time enough to rest. The hardship of the mountain environment required that its people be far from indolent.

For some the hardship was greater than for others. During the late nineteenth century, as increasing population burst the seams of the older settlements, some families moved higher and deeper into the mountains to make their homes at elevations so wintry that only potatoes and oats could be cultivated. Jacobo himself was born in such a place, which for reasons that by then had already been forgotten was called Diamante—diamond.

Diamante was a steep-sided widening of the canyon of the Rio de las Trampas where blue spruce crowded the riverbanks and where the snow lingered long into April. Even by northern New Mexico standards, the Romeros' life there was arduous. In winter when the snow was particularly deep and the family's sheep and goats had no other way of getting food, Jacobo's father, Narciso, would drive a team of horses to break a trail

Photo 46. A dance in Peñasco, January 1943, by John Collier. Courtesy Library of Congress.

to the edge of the forest and there cut down a Douglas fir. The animals fed on the soft foliage for as long as it lasted, and when that tree had been stripped of its needles, Narciso cut down another.

Perhaps because of the demands of living above 8,500 feet, the land seemed fresh, new, and unknown. Sheepherders and cowboys continually made discoveries in the broken and inaccessible back country. Jacobo remembers the day early in the century when one of his older brothers rushed home with the improbable news that he had discovered in the highest part of the sierra a pair of clear green lakes that shone like mirrors. Some of the family thought the boy was lying. The next day Jacobo's father set out to confirm or disprove the discovery for himself. He and several of his older children climbed for half a day through tangled deadfalls and sucking bogs as the canyon steadily narrowed and darkened. One stretch was so steep that they had to climb the land like a tree, using their hands, but at the top they suddenly found themselves in an enormous explosion of space where the wind chased itself in great gusts. At the foot of immense cliffs that rose to the sky they found the lakes, as bottomless as emeralds.

Actually, the lakes by that time had previously been discovered by a number of other people, but in those days news traveled slowly when it traveled at all, and so the Romeros had the privilege of discovering the lakes for themselves.

When an opportunity arose to buy a farm five miles downstream from Diamante in the warmer and larger village of El Valle, Narciso Romero moved his family there. In order to pay for the farm, he began to seek out wage-paying jobs, one of which took him to a section camp south of Raton, New Mexico, where he worked as a laborer for the Atchison, Topeka and Santa Fe Railroad. It took four days to get there, walking all the way and leading a burro that carried his food and gear. Hours were long and wages amounted to no more than a dollar or two a day, but Narciso managed to squeeze from his earnings enough money to buy the first buggy that anyone in El Valle ever owned. It was a marvelous contraption, graceful in every way that ordinary wagons were not. Its wheels were so narrow that they missed the rocks, and its springs made even the washboards of the village roads seem smooth. In spite of the fact that not all the children in the family had shoes and that the youngest were clothed in dresses made of feed sacks, the Romeros counted themselves prosperous.

Often enough, however, that mood was broken by calamity. One such occasion occurred when Jacobo was hauling manure from the barn to spread it on one of the fields. A sudden lurch of the wagon threw him from his slippery perch on a board atop the load. He fell with his leg in the path of one of the burdened wheels, and the horses bolted forward. His leg was crushed. His brothers carried him to the house and summoned his father. There was no question of seeking a doctor's help, for no doctor ever came to the remote mountain villages. It was up to Narciso to manipulate the fracture, align the bones, and finally immobilize the leg in a splint he made from cardboard cylinders that the Romeros used for food storage. Not for twenty days did Jacobo leave the bed, and then only to hobble painfully about on crutches. After two or three months, he tried using the leg and happily found it as straight and sound as the other one. He was lucky; there were plenty of people in those doctorless villages whose lame or crooked limbs testified to a grimmer kind of fortune.

Besides common sense, the only defense of the villagers against pain and sickness was their extensive herbal lore, much of it borrowed from the Pueblos. Herbs were credited with help against every ailment from cancer to the pangs of love. As everywhere else in the world, the medicinal use of herbs, which is still common in northern New Mexico, was a mixture of both real curing and elaborate superstition. One of the most frequently used native plants was oshá (*Ligusticum porteri*), a member of the celery family, which grows several feet tall in the moist meadows of the high country. In various preparations the root of the oshá plant appears to be genuinely effective in soothing a variety of stomach troubles, and the leading herbalist in Santa Fe says that even today he cannot keep enough of the plant in stock. The efficacy of other uses of oshá, however, is harder to credit, including the tradition that a scrap of oshá root in the pocket will protect the bearer from snakebite.

In matters of public health the villagers also had to fend for themselves. Drinking water for both livestock and humans flowed through the same irrigation ditches, and the task of keeping the ditches relatively free of ice during the winter (so that they would not overflow or break) was as important as it was irksome. Since the ditchwater ran through numerous barnyards and pastures, it was hardly as pure as one might have wished, but typhoid, unlike malaria and measles, seems not to have been a serious health problem. As for sanitation facilities, there were none. Not until well after World War I did the privy find acceptance in the mountain villages, and before that not even the crudest latrine was used. In spite of the harsh climate, the only toilets in the villages were the private sides of conveniently located piñon and juniper bushes, and the job of cleaning these sites was relegated to free-roaming pigs and chickens.

Life in the mountain villages was crude and difficult, but these same qualities helped to reinforce the villagers' sense of tradition. They lived as their parents had lived, the same as their parents before them. No one possessing *verguenza* wanted more. Isolated by the great spaces and broken topography of northern New Mexico, the mountain villagers were a unique people. The Anglos called them Mexicans, but the Mexicans did not. The workers who came up from south of the border were known as *Mojados* (literally, *Wets*), for they had dampened their ankles in the droughty Rio Grande. As they labored with native New Mexicans in the sheep camps and mining towns of the West, the Mojados invented a special name for their northern cousins. They called them *Manitos*. The name derived from the New Mexicans' ancient habit of addressing each other as Brother and Sister. Neighbor Juan and his wife María, for instance, were known to their friends as *Hermano*, shortened to "Mano Juan," and "Mana María." These comradely titles underscored the communal character of village life.

One job which the Manitos, because of their permanence in the region, were better able to hold than the Mojados was that of *partidario*. The Mexicans, however, may have been fortunate to miss this opportunity. The New Mexican partido system, as adapted to large-scale commercial sheep ranching, amounted nearly to peonage. Essentially it was a form of sharecropping. A herder contracted with a sheep dealer to rent a certain number of ewes for a period of three to five years, and for every 100 ewes he paid an annual rent, typically of 300 pounds of wool and 20 to 25 lambs.

From Santa Fe north to Antonito, Colorado, Frank Bond and Edward Sargent were the most powerful sheep dealers in the Rio Grande country. Together they managed to acquire grazing rights to so much land, public and private, including land in the newly established National Forests, that many independent sheep ranchers were unable to find sufficient pasture for their own small herds and were compelled to sign on as *partidarios* with Bond or Sargent. The two entrepreneurs, for their part, were happy with this method of recruitment. They preferred their *partidarios* to have sheep of their own, for the herder's animals provided security against the loss of company stock.

In theory, a *partidario* should have earned about fifty lambs and a modest amount of wool for his year's effort on the range, but in practice his gains frequently found their way back to the company treasury. He alone was responsible for bearing the costs and risks of the grazing operation, and he was obliged to outfit himself at the company store, where inflated prices were charged to his account at 10 percent interest. He was

Photo 47. Sheepherder and flock atop Trailriders Wall, Truchas Peaks in background. 13 July 1919. Courtesy Santa Fe National Forest.

further required to buy an appropriate number of rams from the company and to pay for the privilege of using the company's grazing rights at a rate of about twenty-five cents a head for both his and the company's animals. All of his various obligations, not to mention the consequences of losing stock to storms or predators, ensured that he slid deeper in debt to the company every year he worked for it. It was not unusual for a *partidario* to become so mired in obligations to the sheep dealer that he could not expect—and was not expected—ever to extricate himself. His and his fellows' complete financial dependence on the company meant that sheep operators like Frank Bond might profitably write off thousands of dollars of "losses" for bad debts and at the same time be happily assured of an ample supply of cheap, well-trained, and docile labor.

Fortunately the partido system never spread far beyond its original boundaries in the Spanish Southwest, and independent herders seeking straight wages could ride the trains, as Jacobo and many of his neighbors did, to Utah, northern Colorado, and Wyoming, where the sheep industry was booming as vigorously as in New Mexico. None of them ever became rich at their work, but, unlike the *partidarios,* they at least had the satisfaction of not getting poorer. Other villagers, willing to do harder and heavier work for slightly higher pay, found jobs in railroad and logging camps, in beet fields and orchards, and in the mines of Terrero and such Colorado towns as Gilman and Leadville. Altogether, by the end of the 1920s, about one member of every village family was leaving home each year to find work.

The departure of so many workers from the villages was not without precedent. Earlier generations had seen large numbers of men and sometimes women annually venture out to the eastern plains as ciboleros and comancheros. They undertook their dangerous but profitable expeditions in order to supplement the relatively meager harvests of their mountain farms. Now their sons and grandsons sought wage labor for largely the same reason. Most of the village farmlands were only marginally productive. The short mountain growing season, stormy weather, and frequent insufficiency of irrigation water ensured that farming would always be an uncertain and poorly rewarded endeavor.

Worse, the plots of land that most families owned were simply too small to produce a harvest of adequate size. There had never been a tradition of primogeniture in New Mexico, and most farms had been divided repeatedly over successive generations. As a result the average holding of irrigable land was now only a fraction of the fifteen or so acres it had been in the early days of settlement. Not only did the small size of the tracts directly limit the amount of harvest, it also militated against increasing the productivity of the farmer's labor through the use of machinery and the attainment of economies of scale. Elsewhere in America the average size of farms was growing rapidly, as the techniques of industrial production were applied to agriculture. Farmers were becoming businessmen, although often grudgingly, and learning the commercial skills of debt management and capital accumulation. But the Hispanic farmers of the southern Sangres were farming less and less. Although the narrow village fields yielded valuable subsistence crops, they were incapable of producing the one thing that the villagers most desired and most needed: a cash profit.

The Hispanic farming communities of the Rio Abajo, by contrast, possessed abundant farmlands and the capacity to grow cash crops, and they soon found themselves able to participate in the cash economy. The mountain villages, however, did not and, to some extent, still do not. Tucked away in isolated valleys and already stripped of their principal asset, the forests and rangelands of the ejido, the mountain communities offered little to the capitalist world. Moreover, the villagers, bound by their ethic of sharing and *verguenza*, found Anglo notions of progress essentially alien.

Nevertheless, the attractions of manufactured goods ultimately overcame their isolation and their resistance. Cash penetrated the villages the way roots of a tree work their way through rock. As an earlier generation had adopted metal plows

Photo 48. Las Trampas in 1980. Compare with Photo 41, taken sixty-five years earlier. Barbed wire has replaced wattle fences, and tin roofs are more common. Also vegetation on the hillside behind the village is much denser as a result of reduced grazing and fuelwood collection. Most significant, however, is how little the village has changed through the passing years. Photo by Alex Harris.

and other manufactured items, now the Manitos applied their cash wages to the purchase of a range of comforts and labor-saving tools that their parents had never known. Suddenly they could buy spring buggies like Narciso Romero's, cast-iron cookstoves, John Deere farm equipment, more and better mail-order clothes, canned foods—the possibilities were endless.

The blessings of cash were accompanied by a number of other less welcome innovations. Barbed wire came into use in the villages in the decade preceding World War I. The first ambitious farmers who so lacked *verguenza* that they dared to use it provoked considerable consternation among their neighbors. Their spiny fences threatened to destroy the spirit of community trust and communality that had so long prevailed. Although the new wire boded ill for the survival of the old ways, it clearly offered advantages. With barbed wire a man could close off his land so that he alone controlled its use. He could rest and protect it or he could graze it to death, according to his will. Regardless of the injury to tradition, barbed-wire fences soon proliferated in the mountain valleys, and most people, with little delay, came to accept them as part of the increasingly individualized future, the same as wage-paying jobs. The villages even coined a new phrase from the novel sight of livestock tangled in the tricky wire fences (a frequent occurrence since the animals had to learn about the new arrangements too). Now people said that a man was *en el alambre*—in the wire—if he had gone on alcoholic furlough and was not expected to be back to normal for a few days.

Along with the new fences came a new way of life. Village women found that the burden of labor they bore increased. With the men gone for longer and longer periods, the women now frequently assumed responsibility for the farm and livestock, as well as for child care and a myriad of arduous household duties. Eloisa, for instance, supervised the plowing and planting of the fields, the endless chores of irrigation, and the final frenetic effort of harvest all without the help of her husband, who was usually home only for three or four months during the winter. At the same time she also managed to feed, clothe, and keep healthy her thirteen children, all of whom necessarily labored hard in the fields or the house as soon as they were old enough.

As a result of the changes introduced by the cash economy, the security of village life deteriorated. In place of their former self-sufficiency, the villagers were now dependent on either livestock ranching or outside jobs to supply the income that they needed. They soon found that the first of these options was substantially foreclosed by a combination of Forest Service policies intended to improve deteriorated ranges. The second was demolished by the crash of 1929.

The inevitable conflict between the Forest Service and the villagers grew out of two radically different perceptions of the land. Although primarily concerned with the production of timber from the National Forests, the Forest Service was also responsible for protecting the mountain watersheds and rangelands from abuse. Where the villagers saw grasses that could feed their cows and sheep, the Forest Rangers saw the last remaining ground cover of an eroding landscape. Where the villagers saw a vast mountain range that had sustained their forefathers and would sustain them too, the Forest Rangers saw land in the dire final stages of a long-running ecological disaster. Gradually the Forest Service placed restrictions on herd numbers and grazing seasons, and pastoralism, like agriculture in the villages, slowly lost its vigor and declined.

The limitations that the Forest Service imposed were hard on the villagers but not nearly so hard as the Great Depression itself. To their great misfortune, the final collapse of their pastoral subsistence economy coincided with the economic collapse of the entire nation. From 1929 onward few villagers were able to find outside employment, and the flow of dollars from the mines, sawmills, and sheep camps slowed to a trickle. The burden of poverty increased drastically in the villages, and it was hardly salubrious. Malnutrition, especially among children, became a serious and widespread problem, and in 1935 a Taos County nurse reported that nearly a third of the school children in the Peñasco Valley were seriously undernourished. Only the massive aid programs of the federal government prevented widespread starvation, and by the middle 1930s between 60 and 70 percent of northern New Mexicans were on relief. Incredibly, in the span of a single generation the mountain villages had fallen from self-sufficiency to total dependence on government subsidy.

Today the people of northern New Mexico still lack a viable economic base, and they still depend heavily on government assistance. The small farms of the villages produce some hay and beef, both of which are readily converted into cash, but only a handful of villagers earn a substantial part of their living from agriculture. In order to make ends meet, many villagers find jobs with the Forest Service, the school system, or the construction trades, but there are never enough jobs to go around. The Hispanos of the southern Sangres are among the poorest people in the United States. In 1970 nearly 60 percent of the residents of Mora County lived in what the federal government officially defined as poverty. For Rio Arriba, San Miguel, and Taos counties, which are also predominantly Hispanic, the figure ranged between 34 and 40 percent. Nationally fewer than 11 percent of American families belonged to that category.

Because economic prospects were so bleak in the years after World War II, vast numbers of young people left the villages and moved to such urban centers as Albuquerque, Denver, and Los Angeles. They left behind scores of boarded-up houses and decaying barns, and their departure depleted the villages of both skills and leadership. Today there are scarcely enough people left in some villages to maintain the churches and the irrigation ditches. The old village customs and ceremonies, once so rich, suffer similar neglect. Truchas and Chimayo were obliged to combine the celebration of their saints' days in order to reduce expenses and save the priest's time. In El Valle, as in most villages, the santo is no longer carried forth to bless the fields and the ditches, but is simply given a turn around the church and then put back in his niche. His services apparently are not required now that the people have ceased to depend in any substantial way upon the land.

The exodus from the mountain villages was never a joyful one, for there was nothing in the cities to replace the beauty and the strong community ties of the mountain world. Unfortunately, the conditions that caused the exodus have still not changed, and few young people who stay in the villages escape chronic unemployment. Understandably, many of them are bitter about their fate. They know that their patrimony of land grants was taken from them by Anglo outsiders, and many

Photo 49. Jacobo and Eloisa Romero, 1981. Photo by Alex Harris.

of them further blame their poverty on other unfriendly Anglo influences and institutions, including the Forest Service. Beginning in the late 1960s they began to espouse the political and cultural goals of Mexican-Americans throughout the Southwest and to flex their political muscle under the slogan of *Chicano Power*. The new consciousness led to the establishment of various village clinics and cooperative economic ventures, as well as to numerous stormy political confrontations. Whether it will lead further, as many hope, to new social and economic vitality for northern New Mexico remains to be seen.

The rhetoric of national Mexican-American politics, meanwhile, should not obscure the unique heritage of the mountain people. Although the Hispanos of the southern Sangres belong to the greater brotherhood of *chicanismo*, they also belong fundamentally and locally to themselves. Eloisa Romero, in her bright kitchen where a woodstove pops and grumbles beside a modern gas range, pauses when asked about the problem of changing names. "I don't know what we are supposed to call ourselves now," she says. "I never heard that word *chicana* when I was a girl, so I guess I cannot be one now. What do you think, Jacobo?"

The old man shrugs. "If you ask Juan de Dios," he says, "he will tell you we are Spanish."

"But we are not Spanish; the people in Spain are Spanish, not us."

"Well, we have to be something, and so I guess I will be a Mexican."

"No, you are no Mexican either, old man. A Mexican is one of those boys who comes here from Juárez or Chihuahua to make adobes."

"Well then, if I am not a Mexican, and I am not Spanish," he says, savoring the riddle, "what am I?"

She puts her hands on her hips and appraises him with a cattle-buyer's squint. "I am not too sure what you are, old man, but I suppose you are probably some kind of an old Manito."

14

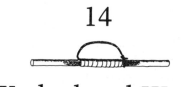

Washed and Worn

Don't get excited. You know lots of times the only difference between a good sheepherder and a bad one is how much range he can steal.

—Frank Bond

THE OLD-TIMERS SAY that if there is one way in which the old days really were better than the present, it is that they were wetter. They talk about dry arroyos that once ran with water and rangelands that used to be green and lush. The winter snowpacks that they remember were arctically deep, and spring runoff brimmed in the irrigation ditches until well into July. The proof of the wetness of those days, they say, can be seen anywhere in the mountains. At the edge of most villages lie abandoned fields that once were irrigated and sown with grain. You see them best from the air or from atop one of the high peaks. Studded with young pine, chamisa, and juniper, they stand out as bands of gray-green, intermediate between the light color of the village hayfields and the dark of the surrounding forest. It is because the winters of recent decades have been puny and ungenerous, so the story goes, that those fields were abandoned, and the range grasses dried up, and a hundred other ills descended on the villages to sap their vitality.

There is truth in that account, but it is not the whole truth. According to meteorological records, which have been reliably kept in New Mexico since the late nineteenth century, and tree-ring data which give a picture of the climate before then, the early decades of the twentieth century were indeed a period of abundant moisture. In fact, the years from about 1905 to 1920 appear to have been the wettest fifteen years in the last two and a half centuries. Many older New Mexicans remember this relatively brief period as typical of all of the "good old days," even though no period could be less so. Before 1905 and after 1920 precipitation in the region was more or less the same as it is today. But the old-timers who generalize about a past when the stock ranges were always greener and the creeks and ditches fuller than today are not altogether wrong. The land has changed, drastically and dramatically, and no corner of the mountain country—or of New Mexico—has been untouched. The main agent of change, however, has been not the weather, but man himself.

If the soils and streams of New Mexico could speak, they might tell a story of destruction more dismal than anything the old folks relate. They might begin by describing what the good old days did to the Galisteo valley, twenty miles south of Santa Fe. In 1849 a wagon train bound for California paused there to fatten its stock on the valley's grasses, the ranges nearer Santa Fe having by then been denuded by more than a century of continuous use. One member of the emigrant train, a Quaker named C. E. Pancoast, noted in his journal that around the town of Galisteo there were "many acres, planted with Corn and other Vegetables, and irrigated." If he returned today, however, Pancoast would surely fail to recognize the town. Irrigation has been abandoned since 1926, and the "large Plain of grass" where the wagon train's oxen and horses grazed (and which was also the principal pasture for the stock of the Army's Fort Marcy in Santa Fe) is now deeply cut by arroyos and all but barren. Pancoast wrote that on his way from camp into the town of Galisteo, "I had to walk a plank across a Creek." The Quaker mentioned the plank because after a visit to a cantina, where he enjoyed several drinks of turpentine-flavored aguardiente, he found that the plank had become so dangerously twisted and unstable that he was hard put to cross it at all. No such danger now faces visitors to Galisteo and not just because turpentine is left out of the whiskey. The Rio Galisteo, which in all likelihood is the creek that Pancoast crossed, can no longer be spanned by any single board. In the vicinity of the old plaza, its dry gravel bed has grown a dozen feet deep and up to two hundred feet wide.

The story of the Rio Puerco, which flows southward from Cuba to the Rio Grande on the west side of the Jemez Range, is even more tragic. In 1880 some ten thousand acres of the upper Puerco valley between Cuba and Casa Salazar were under irrigation, and its fertile but fragile alluvial soils supported a half-dozen small subsistence farming communities. Forty-five years later, however, fewer than three thousand acres were still in production, and agriculture in the upper valley was in a shambles. Running with a wildness unknown in former years, the river had washed out or silted in the headgates of its irrigation ditches; it had broken through diversion dams and scoured its channel so deeply that its waters were as inaccessible to local farmers as though they flowed on the far side of the moon. By 1951 irrigation downstream from Cuba was impossible, and the villagers abandoned their fields and homes. Upstream from Cuba thousands of additional acres of irrigated land had also been lost to erosion, and hundreds of arroyos radiated outward from the valley bottoms, cutting through the uncultivated rangeland, virtually to the tops of the mountains. All told, nearly a billion cubic meters of earth eroded from the Puerco watershed since 1885, rendering it, for all practical purposes, a desert.

Certainly a fundamental cause for the rampant erosion in the Puerco Basin was the unusual character of its soils, which combine weak structure with exceedingly high solubility. Running water easily carried them away. But while the cause of erosion was partly geologic, the major responsibility for it belonged to the settlers of the Rio Puerco, who stocked the ranges of the watershed with too many sheep, goats, cattle, and horses. The hungry animals stripped the vegetation from the land, robbing the soil of protection, and with their numbers they created trails that channeled rainwater like gutters and grew into deep gashes in the earth. In all of North America there is no place that illustrates better the destruction that can result from failure to adapt to the limitations of the environment.

The man-caused changes that demolished agriculture in the Rio Puerco and Rio Galisteo watersheds also dried up and eroded ranges throughout New Mexico and seriously undermined the vitality of farming in the mountains. Because the changes accumulated slowly, year by year, most farmers and ranchers attributed them to the weather. Obedient to the necessity of making a living, of extracting the greatest short-term benefit from field and prairie, these sharp observers of storm and drought were almost universally unwilling to admit their own role in provoking the changes that beset them. It remained for a new breed of scientifically trained resource managers to begin the campaign for environmental responsibility that continues to the present day. In 1927, J. Stokley Ligon, a biologist for the New Mexico Department of Game and Fish, concisely summed up the argument that by then had divided resource managers and stock raisers for a generation:

In recent years ranchmen have been complaining about the country becoming more arid and contending there has not been as much rainfall as formerly, while in truth, the amount of moisture has been about the same, but the denuded ranges can no longer stand abuse. Man alone is at fault and not the elements.

Ligon may have too categorically denied the possibility of significant climatic change, but he did not exaggerate the dire condition of New Mexico's stock ranges. Overgrazing had been epidemic throughout all the western states for more than fifty years and had caused the productivity of the region's grasslands to decline by more than 50 percent. In spite of the protestations of ranchers to the contrary, the role that climate may have played in the loss was certainly minor; more important, its practical significance was nil. Regardless of how much rain "ought" to fall, ranchers should have adjusted the size of their herds and their grazing practices to the amount of rain that did fall. By failing to adapt to the environment, they injured both themselves and the land.

The cattle industry that was born in New Mexico at Fort Union and nourished both by comancheros and by Texans who plied the Goodnight-Loving Trail had grown furiously following the completion of the Atchison, Topeka and Santa Fe Railroad to Las Vegas in 1879. In 1880 there were only 137,000 beef cattle in the territory; four years later there were a million, and four years after that, a million and a quarter, many of which were brought to New Mexico by Texans fleeing the effects of overgrazing in their own state.

Although the greatest concentration of cattle was on the territory's northeastern plains, large numbers of them also grazed in the southern Sangres. They were brought there from every village and flatland in the shadow of the mountains. Stockmen usually drove their herds into the high country in early May, as soon as plowing was done and snow had melted in the mountain valleys. The Hispanic ranchers of Le Doux and Rociada, who took their cattle to the headwaters of the Pecos, faced one of the most difficult tasks of trail driving. Their only access to the high country was a pair of treacherously steep and rocky trails that led up the wall of the East Divide. In years when the snowpack was particularly deep, they cut steps and shoveled a passage through the monumental drifts in order to push their cows to the pastures beyond.

At that time of year the village hayfields had scarcely turned green, but the cattlemen were either unable or unwilling to acknowledge that the high-altitude mountain grasses, being still less advanced, had only begun to awaken from their winter sleep and could not be grazed without damage. Worse, the grasses never had respite enough from grazing to flower and reproduce. All summer long the cattle cropped their seedless stems until the advancing snows of autumn drove the livestock—and the cowboys—to warmer and lower elevations.

Sheep presented an even graver problem. As Lieutenant Carpenter of the Wheeler Surveys had predicted, New Mexico did become the preeminent sheep country of the West. The territory's sheep population soared from 619,000 in 1870 to a peak of nearly 5 million in the late 1880s, and it did not drop below 3.5 million until well past the turn of the century. Sheep were usually herded in bands of one or two thousand, and as they were driven from place to place, their sharp, tiny hooves did even more damage than their mouths. They grazed slopes much steeper than those a cow would climb, and heavily used slopes became contoured with so many dozens of skinny trails that they began to look like corduroy. Damage was especially great in the mountain bedding grounds, which at higher altitudes were limited mainly to the cirques. With thousands of mouths and four times as many hooves attacking the plant cover night after night, the mountains were all but stripped of their native vegetation.

Photo 50. Ranchers from Rociada bring their herd over the East Divide to summer range in the Pecos Wilderness, 1976. Hermit's Peak in distance. Author photo.

Photo 51. A New Mexican sheepherder, c. 1935, by T. Harmon Parkhurst. Courtesy Museum of New Mexico.

Elsewhere in New Mexico the boom years of the 1870s and 1880s brought on a period of harsh and often bloody rivalry between sheep and cattle operators, but in the mountains competition for rangeland was mild when it existed at all. It was to the good fortune of the meat and wool growers and the great misfortune of the land that cattle and sheep favored different kinds of mountain rangelands. Sheep liked the open spaces, short grasses, and tangy forbs of the high divides and found the deep bunchgrasses of the lower parks relatively unpalatable. Cattle, on the other hand, relished the bunchgrasses and avoided the windy, unprotected alpine slopes. As a result, every square foot of the mountain rangeland was put to full use, and the use continued without a break from snowmelt in the spring to the first white storms of October.

On the strength of their biological reserves the rangelands

were able to maintain their productivity for a number of years, but their ability to do so, unlike the willingness of ranchers to take the greatest possible number of livestock to the mountains, was finite. Slowly the relentless teeth and sharp hooves took their toll. With no opportunity to reproduce, the deeply rooted climax grasses, including alpine timothy on the summits and Thurber fescue in the parks, died back. They were replaced by less productive grasses and by unpalatable forbs like yarrow and fleabane, which gave comparatively little protection to the soil. Once the deterioration began, it advanced rapidly, but hard use of the mountains continued with no loss of intensity. In August of 1898 the *Monthly Bulletin* of the National Wool Growers' Association reported that overgrazing on the Pecos Forest Reserve had progressed to the point that the hungry sheep were even stripping the trees of their foliage.

One of the worst aspects of the "mining" of the rangelands was that it destroyed the capacity of the land both to absorb water and to store it. As plants died, more and more bare spots opened in the spaces between their surviving neighbors. And these spaces were truly bare. There was no surplus forage to fall and replenish the mulch of decaying organic matter that had previously protected the soil, both as a sponge to soak up moisture and as a blanket to hold it in. Now when the rain fell it landed directly on the dirt, loosening soil structure with its considerable kinetic energy and scattering soil particles with a splash. Especially on steep slopes, the unprotected soil could absorb little of the rainfall before rain and dirt together began to run in chocolate rivulets. And when the sun came out again, it quickly parched the naked ground, leaving little moisture to sustain the plants that remained. A further problem greatly reducing the vitality of soils and vegetation was the sheer weight of the livestock. Their almost continuous trampling of the grasslands, especially when the ground was soft following spring snowmelt, compacted and hardened the mountain soils, reducing further their capacity to absorb the precipitation of future storms.

It must be noted that the weather played an important supporting role in the depletion of New Mexico's range resources.

Photo 52. *San Luis Diversion Dam, 19 July 1945. Built in 1935 to raise the level of the Rio Puerco sufficiently to put water in village irrigation ditches. Courtesy Santa Fe National Forest.*

Between 1820 and 1835, when Indian depredations were at a minimum, New Mexico's available sheep ranges were stocked nearly as heavily as they later were during the 1880s and 1890s, but with a few exceptions, they evidently survived that period of heavy use with little permanent damage, probably because, as tree-ring data indicate, New Mexico then enjoyed slightly greater than average rainfall. Unfortunately the second half of the century was not similarly blessed. The years of the stock-growing boom (and not the hard years after it) were somewhat drier than usual, and they seem also to have featured more than their share of virulent, flooding thundershowers, which, even under the best conditions, cause some erosion; under the deplorable conditions that were then almost universal, they brought disaster.

Arroyos—gullies that run intermittently with storm runoff or snowmelt—had not, until then, been especially common in New Mexico, but because of the weakened condition of the land, heavy showers and flash floods soon made them one of the commonest features of the Southwestern landscape. Arroyos further helped to wreck the health of the land by cutting into and lowering the water table, and they added to social and economic woes by destroying roads, trails, and irrigation systems. Roads and trails, in fact, greatly facilitated the spread of arroyos since every wagon rut or bare horsepath was an incipient channel that the weather might easily enlarge. Along the Camino Real between Santa Fe and Taos, parts of which are roughly followed today by New Mexico State Roads 75 and 76, almost every grade of the old dirt highway is marked by some kind of gully.

The stress of overgrazing radically changed the vegetation of rangelands throughout New Mexico. Climax perennial grasses, grazed literally to death, gave up their territory to quickly sprouting annuals and weedy forbs and shrubs, as well as other, less palatable perennials. Because of the loss of both dead and living vegetation and because of trampling and compaction, the microclimate of most rangeland soils became increasingly hot and arid. Fewer plants shaded the surface of the soil, and there was less mulch to slow evaporation. Consequently the range plants were obliged either to adapt to a desiccated environment or to surrender their place to other plants better adapted to the new conditions. Grasses like western wheatgrass and blue grama, which are widespread in the piñon-juniper woodland, began to appear higher up in the openings of ponderosa pine forests, where they competed successfully with drought-stricken grasses formerly dominant in that ecozone. The migration of flora from lower to higher altitudes and from south-facing to north-facing exposures is an important indicator of lands that are losing or have lost their native vigor. Where it occurs, it bespeaks a tragic loss of productivity.

As a result of overgrazing, big sagebrush (*Artemisia tridentata*) now dominates the vegetation of vast tracts that were grassland or open savannah a hundred years ago. Overgrazing also nearly eradicated a half-dozen species of cool season grasses that once formed a major component of the vegetation in the piñon-juniper woodland. Cool-season grasses utilize winter and spring moisture and carry out their growth and reproductive cycles before the heat of summer arrives. Because they were the first green forage plants to appear after the long dormancy of winter, cool-season grasses were especially valuable to stockmen and their herds. But for the same reason, because virtually no other green forage was available from April to June, the cool-season grasses were grazed almost to extinction. Still today the relative scarcity of cool season grasses in the piñon-juniper ecozone severely limits stock raising in the northern part of the state.

Higher up in the mountains the effects of overgrazing were no less pervasive. Today there is not a single meadow, cirque, or bald divide in the southern Sangres where the original, native flora of range plants remains intact. In fact, so thoroughly did domestic animals rework the flora of the high country that in many cases it is impossible to say with certainty what the original flora actually was. In what must have been a dizzying botanical revolution, native species that could not withstand heavy use were replaced by invaders coming not only from lower elevations, but also from distant parts of the world. Sheep fescue (*Festuca ovina*), for instance, is originally native to Eurasia, but its seeds came to New Mexico in the fleece and droppings of domestic sheep, possibly as early as Oñate's day. Today it is the most common element in most of the Sangres' alpine grasslands. Similarly Kentucky bluegrass (*Poa pratensis*), another Old World immigrant, is a major constituent of the high country's subalpine parks, where it commonly fills the spaces between clumps of native bunchgrasses like Thurber fescue.

Although grasses like sheep fescue and Kentucky bluegrass are recent additions to the vegetation of the Sangres, it is a good thing that these and other durable exotic species were available. Without their vigorous colonization of the tired mountain rangelands, even more of the high country soils would have washed from the slopes. Nonetheless, the ubiquitous presence in the mountains of nonindigenous species is a reminder that, at least with respect to the Pecos and other high country wildernesses, the legislative definition of wilderness as "an area where the earth and its community of life are untrammeled by man" will be forever unattainable. The notion that wilderness bears no human imprint tends to blind people to the complex and stormy heritage of these rugged lands.

Overgrazing was not the only woe afflicting the mountain country. The mountain forests also suffered destruction on a large scale, often with severe damage to soils and watersheds. Natural wildfires burned many forests, while stockmen intentionally set others ablaze in order to make more land available for grazing. Loggers, meanwhile, cut the timber from tens of thousands of acres, with no thought for regeneration, in

Photo 53. A tie-cutting camp of the Santa Barbara Tie and Pole Company, c. 1909. Courtesy Santa Fe National Forest.

order to satisfy the territory's ferocious appetite for railroad ties, mine props, and sawtimber.

The largest timber operation in the southern Sangres was undertaken, strangely enough, on account of the Panama Canal, which began operating in 1914. By using the canal, shipping lines promised to undercut the freight rates charged by railroads like the Atchison, Topeka and Santa Fe. In order to remain competitive the AT&SF elected to increase the speed and efficiency and thereby lower the costs of its operation by laying a second set of tracks across New Mexico and Arizona. One of the principal requisites of the construction was the procurement of an abundant source of ties, for in building its first track, the AT&SF had already consumed the timber resources of the lands adjacent to its right-of-way.

At least sixteen million new ties were needed, and in 1907 an audacious Yankee by the name of A. B. McGaffey finally located them in the high spruce forests surrounding Jicarita Peak at the north end of the present Pecos Wilderness. With little difficulty he purchased 24,750 acres of the Santa Barbara Land Grant, which not long before had been alienated from its Hispanic owners by means of a suit for partition, and quickly the Santa Barbara Tie and Pole Company was born. A short while later McGaffey also purchased the standing timber on 41,000 acres of the adjacent Mora Grant, and by 1909 his operation was in full swing.

For seventeen years, until it ceased operations in 1926, the Santa Barbara Tie and Pole Company was a marvel of both rapaciousness and ingenuity. Its men cut timber all the way to timberline, taking every tree that would make a tie. Portable sawmills were set up at elevations as high as eleven thousand feet, and the cut ties were shot down the mountain sides in narrow wooden flumes whose ruins lie rotting in the forests today. Lower down in the Santa Barbara Canyon, where the company built the logging camp of Hodges, the forest trees were so large that hauling them to a suitably large sawmill proved to be a major problem. To solve it, the company freighted a locomotive and five miles of track into the high country, hauling all of the steel and heavy machinery in mule-drawn wagons from the rail siding at Embudo. McGaffey and company then laid their rails up the side canyons and over the mesa tops of the Santa Barbara watershed, and before long their narrow-gauge shunt line was rolling. All the branches of the tiny railroad connected to the mouth of the main canyon, where McGaffey built a steam-powered sawmill with a capacity of a thousand ties a day.

At its peak the Santa Barbara Tie and Pole Company employed some three hundred men whom it paid $1.75 a day. Many other men from the nearby villages, including those of the Las Trampas Grant, earned their livelihood by cutting ties from their own land and selling them to the company. A squared sixteen-foot log brought 25¢, and an eight-footer brought a dime. Depending upon their skill with broadaxes and the stoutness of their draft horses, two men might deliver $5 or $6 worth of ties a day.

High-altitude flumes and dinky railroad notwithstanding, the most impressive accomplishment of the Santa Barbara operation was the manner in which the ties were transported from the mountain forests to the AT&SF creosoting and fitting plant in Albuquerque. All winter the cut ties were stockpiled at Tres Ritos on the Rio del Pueblo and at Hodges and Peñasco on the Rio Santa Barbara. Movement of the ties was restricted to two or three weeks in late April and early May when the streams crested with snowmelt and the ties could be thrown in, by the hundreds of thousands, to smash down banging and jamming through narrow canyons to the river's confluence with the Rio Grande at Embudo. From Embudo the ties were rafted two dozen miles downstream, nearly to Cochiti Pueblo, where they were boomed out of the water and hauled on a spur railroad to a creosoting plant in Albuquerque.

Today one looks at the Rio Grande, especially where it crashes through the jumbled rapids of White Rock Canyon downstream from Española, and it is hard to believe that the river ever carried such enormous loads of timber. Even more incredible, however, is the idea that the skinny Rio Pueblo and Rio Santa Barbara could have floated those millions of tons of wood, for neither is ample enough, at least today, to launch

Photo 54. Tie boom above Cochiti. Photo by T. Harmon Parkhurst, probably 1916. Courtesy Santa Fe National Forest.

a canoe. Admittedly, McGaffey's tie drivers did not rely solely upon the natural flow of the rivers. They built crib dams with which to store a temporary "head" of water. When time came to drive the ties, they dynamited the dams and released a surge of water that carried off the ties. Today, however, even with scores of crib dams, log driving on the scale achieved by Santa Barbara Tie and Pole would be impossible.

One reason for the company's success was that the period during which it was most active enjoyed higher than average precipitation. But that may not have been the most important reason. Recent studies indicate that streams on the west slope of the Sangres today provide only about half the water they did in the 1940s and that climate alone does not account for the change in flow. Rather, it appears that changes in the vegetation of the higher mountain slopes were responsible for the high rate of streamflow in the 1940s, and it is not unreasonable to speculate that their condition thirty years earlier would have yielded even greater amounts of runoff. The overgrazed ranges of the high country then shed water that, if healthier, they would have absorbed, and the high altitude forests also tended to give up their moisture rapidly. Ordinarily the spruce-fir forests shade and protect the winter snowpack, in a sense refrigerating it and releasing its meltwater slowly and gradually. But under the conditions that prevailed early in the century, so much of the high country had been burned or logged that the snowpack melted quickly under the spring sun and swelled the rivers into torrents.

Because of the condition of high country ranges and forests, the runoff that carried ties to the Rio Grande was a violent event. Since so much of the snowpack was exposed directly to the sun, the river waters crested earlier and much higher, and then subsided more abruptly than they do now. The spring runoff also involved more water for a given amount of precipitation since with less vegetation on the mountain slopes, less water was transpired by plants. Because the conditions that prevailed along the Rio Santa Barbara also characterized virtually every other watershed in the southern Sangres and the rest of the region, the enormous amount of runoff from snowmelt and major storms soon became a subject of major concern. Especially when tropical depressions laden with moisture from the Gulf of Mexico dumped heavy rain for days at a time on the high country, the raging streams washed out diversion dams, scoured channels, and clogged irrigation headgates and ditches with silt. Although agriculture probably expanded in the southern Sangres until the turn of the century or a little after, cultivation of much of the newly cleared, marginal farmland soon had to be abandoned due to the difficulty of bringing to it an adequate supply of water.

As the old-timers say, the old days were wetter, at least some of the time, than the present. But the arroyos that flowed like streams and the streams that flowed like rivers should have been interpreted not as signs of the land's fruitfulness but as symptoms of a grievous environmental disease. The desiccation of the rangeland for lack of mulch and living plant cover was another such symptom, and so was the channel cutting that turned the gardens of the Puerco and Galisteo valleys into sand. More dramatic than any other indication of the land's illness, however, was the behavior of the Rio Grande.

Muddy and thick, the waters of northern New Mexico carried the mountain soils downstream. Upon reaching the flatter gradient of the Rio Grande, the waters slowed and dropped their load of sediment, and the bed of the Rio Grande gradually rose. By itself, the aggradation of the river was a significant problem for it caused many thousands of acres of previously productive bottomland to become waterlogged and useless for farming, but far more serious was the likelihood that in times of heavy flow the river might too easily pass its newly inadequate banks and unleash a flood of disastrous proportions. And that, in fact, is exactly what happened in 1884, 1904–5, 1912, 1920, 1929, and 1941.

One of the earliest and most important missions of the U.S. Forest Service was to control this dangerous and destructive situation by limiting grazing on public lands. Unfortunately, just when the rangers were beginning to make real progress, the United States joined World War I. In the patriotic fervor of the day it was widely believed that the nation should produce every possible ounce of meat and wool for the sake of the war effort. The Secretary of Agriculture accordingly directed the Forest Service to increase, not decrease, the number of animals permitted to graze in the National Forests. As a result, restrictions on herd size and grazing seasons, which had required years to establish, were lifted, and the intolerably bad state of the rangelands was allowed, even encouraged, to get worse. Sadly, this new sacrifice failed to achieve its original, limited goal. By the time the increase in permitted grazing produced the desired result of bringing more meat and wool to market, the war was over and the livestock market had shrunk back to its prewar size. The overproduction of animals now caused prices to plummet, and many ranchers, having gambled for a boom, lost everything they had.

And the land, meanwhile, was still losing soil at a calamitous rate. By 1936 Forest Service analysts estimated that at least three-quarters of the watershed of the Rio Grande in northern New Mexico and southern Colorado was suffering serious, accelerated erosion. Unfortunately for the mountain villagers who had lost their jobs in the Great Depression and who now turned to the land for subsistence, the land was as poor as they. Because of a succession of dry summers as well as the depleted condition of the land, it was not unusual for village livestock to starve to death during the winters of the middle 1930s. In 1934, for instance, twenty-five cattle and horses starved in Ojo Sarco alone. That same year in Truchas the villagers fed cactus to their animals to keep them alive, and although such an adaptation to hard times may be relatively common in southern New Mexico or Texas, it certainly was unusual at 8,000 feet among the pines of the southern Sangres.

Perhaps it is only coincidence that the evils of land abuse in the mountain world became most glaringly apparent in the Great Depression. Unfortunately, the two events did not end as simultaneously as they had begun. When the depression eased, the land was still tired and sick. Even today the healing of the land progresses slowly, and the damage is far from corrected.

Some historians have argued that the mountain villagers lived in harmony with their environment as long as their simple subsistence economy remained intact. According to this line of thinking, the environmental destruction that characterized the late nineteenth and early twentieth centuries was a by-product of the cash economy and the consequent commercialization of farming and stockraising. There is undoubtedly a measure of truth to this, for the deterioration of mountain rangelands could not have been so rapid or pervasive had there not been ready markets for the enormous numbers of sheep and cattle that actually carried out the destruction. Nevertheless, it would certainly be wrong to laud the early-day subsistence villagers as guardians of a careful balance in their use of the land. The severely eroded hills surrounding the older mountain villages testify eloquently enough that land abuse predated the advent of the cash economy. Commercial markets may have been responsible for the ultimate epidemic of overgrazing, but they did not invent the disease.

Northern New Mexico's experience with land abuse is not unique to cash economies, to modern times, or to North America. Overgrazing and deforestation are two of the most ancient plagues of mankind. Because of them, most of the Middle East today is a desert, and the productivity of North Africa, Greece, the Yangtze watershed, the Himalayas, and many other regions is only a fraction of what it once was. In 1978 the United Nations estimated that fourteen million acres of land worldwide were lost annually to the man-caused spreading of the world's deserts. Indeed, over the ages humankind has so chronically misused the land that sustains it, and continues to mis-

Photo 55. The Rio Grande at high water of the May 1941 flood, viewed from the dike protecting Española. The Sangre de Cristo Range rises in the distance. Photo by Rex King. Courtesy Santa Fe National Forest.

use it so persistently today, that one is hard pressed to find reason to believe that a harmonious relationship between a society and its environment is possible. Yet such a relationship must be sought—and found. If history teaches one lesson unequivocally, it is that no society or culture can long survive if it squanders the ecological base on which it is built. The story of northern New Mexico in the twentieth century is in large measure the story of a society struggling to find the necessary balance in its use of the land. Since its creation, the U.S. Forest Service, the subject of the following chapter, has been at the center of that struggle.

15

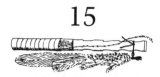

Bully Boys
and Bureaucrats

*I hand you herewith enclosed copy of letter
addressed from Willis, New Mexico, and dated
January 18th, and signed Dr. Wm. Sparks, in
which you will find this term "low down white
livered liar," referring to myself, and further along
in the letter the term "dirty bastards," referring to
all people whom he had seen from Kentucky,
except Wilhoit, who was my predecessor in office.*
　　　　—R. C. McClure, Supervisor of the
　　　　　　Pecos River Forest Reserve, and
　　　　　　a Kentuckian, to the Postmaster
　　　　　　General, 1901.

LAND ABUSE FOLLOWED the conquest of the West like a shadow. Again and again, the same errors of exploitation were committed in every state and territory. As early as the Civil War years, cattle and sheep starved by the thousands throughout southern and central California, having so overgrazed the rangelands that the native vegetation of perennial grasses was permanently destroyed. Although state officials fully recognized that overstocking as much as drought was responsible for the disaster, California's hard-won lesson was lost on Arizona, Wyoming, Montana, and other cattle-growing regions. Through the late 1800s similar die-offs of livestock, caused always by a combination of overstocking, range deterioration, and bad weather, became common throughout the West. And in parallel fashion, westerners rapidly cut down their mountain forests in order to build towns, prop mine tunnels, and lay thousands of miles of railroad track. Amid the frenzy of expansion, few people gave thought to the regeneration of forests and grasslands or to the protection of the watersheds to which they belonged.

Traditional wisdom explains the rapaciousness of western settlement as the result of a simple misperception: Americans wrongly but understandably believed that the resources of their continent were inexhaustible, and so they blundered their way through a century or more of deforestation and overgrazing. This is true up to a point, but it in no way explains why environmental destruction in most cases continued long after the resources involved were recognized to be finite. The reason for the persistence of land abuse in the West was that everyone had strong economic incentives to make fast use of the grass and trees, and no one had reason to hold back. Self-restraint was self-punishment: it inevitably allowed someone else to reap the harvest, and the riches, first. Until the government stepped into the business of land management, the western commons were harshly abused, both by those who cared nothing for the land and by those who loved it.

In spite of the great need for intervention, the idea that the federal government should regulate the use of the millions of acres of unsettled western lands had few supporters until the end of the nineteenth century, and even then it remained intensely controversial. The establishment in 1872 of Yellowstone National Park, the nation's first, was a victory not so much for conservation as for tourism. The first proposal for the park came from an agent for the Northern Pacific Railroad. This farsighted individual probably calculated that more people would visit the park if its geysers and other natural curiosities were kept out of the hands of private speculators. And more visitors would surely mean more fares for the Northern Pacific.

Not until 1890 with the creation of Yosemite, King's Canyon, and Sequoia National Parks did Congress begin to assign the government the task of protecting lands in the public domain in order to preserve their resources and natural integrity. A few months later, in the spring of 1891, Congress again moved in the direction of conservation, albeit almost unconsciously, by passing the Act of March 3, 1891, "For the repeal of the Timber and Stone Act and for other purposes." Attached to that bill was a seemingly minor, undebated amendment that empowered the president to withdraw lands from the public domain and to set them aside as federal forest reserves. With little delay President Benjamin Harrison acted on his new prerogative and extended the government's protection to thirteen million acres of western mountain country. One of the reserves created in this first generation of what later would become National Forests was the Pecos River Forest Reserve, comprising most of what is now the Pecos Wilderness. This was the first national recognition of the natural beauty and resource value—especially to downstream water users—of the Pecos High Country.

At first the new forest reserves were indistinguishable from national parks. The legislative amendment that made them possible had specified neither their purpose nor how they were to be managed. Some conservationists felt they should be preserved strictly for the sake of their pristine character and scenic beauty, while others, notably an innovative forester named Gifford Pinchot, advocated using them for steady, long-term production of timber and other natural resources. For a number of years, however, the debate over the future of the forest reserves was entirely academic, for Congress had voted no funds for administering them. In the meantime, the intensive exploitation that federal protection was intended to control was allowed to continue unchecked.

Finally in 1897 Pinchot and his camp won passage of the Forest Management Act, which instructed the Secretary of the Interior to begin active management of the Forest Reserve System, which by then had swelled to forty million acres. Unfortunately, the Secretary delegated his new responsibility to the General Land Office which was then one of the most corrupt and inefficient bureaus in the government, and nowhere did its administration of the forest reserves begin auspiciously, least of all in the Pecos. The first supervisor of the Pecos River Forest Reserve was a political appointee from Kentucky named James B. Wilhoit, who, after drawing his pay for a number of months, finally decided to undertake an inspection of his new domain in July of 1899. Being unused to western ways and western horses, Supervisor Wilhoit packed a large black umbrella instead of the conventional slicker for protection against the elements, and as he rode north along the windy crest of Hamilton Mesa into the heart of the high country, a thunderstorm began to spatter him. Without dismounting, Wilhoit commenced to deploy his cumbersome umbrella, but the horse he had rented that morning for a dollar a day took exception to the black flapping creature that materialized within chewing distance of its ears, and began to pitch and yaw with all the passion in its bones. Supervisor Wilhoit was forthwith launched into space, and when he again rejoined terra firma in the middle of a snowberry bush, the first inspection tour of the Pecos River Forest Reserve officially and abruptly ended.

The affairs of the Pecos Reserve improved little under Wilhoit's successor, R. C. McClure, who soon after his arrival in New Mexico became embroiled in a feud with Dr. William Sparks of Willis (later Cowles, New Mexico). A country doctor who patched up animals as well as humans, Sparks operated a small sawmill when he first settled in the Pecos country, but then suffered the misfortune of losing an arm in his planing machinery. Later he became postmaster of Willis, and it was from that position of relative authority that he began to accuse McClure of myriad sins including nepotism, embezzlement, and the seduction of a widow on government time. McClure in turn denounced Sparks as a "blatant Bryan Free Silver Democrat and a calamity howler" to all the influential Republicans he knew. The feud with Sparks helped ensure that throughout his tenure McClure concentrated more on protecting his own political and bureaucratic future than on restoring the vitality of the Forest Reserve. Like many of his counterparts on other reserves, his main achievement in government service was his initial success in obtaining a job.

The Forest Reserves were among the most treasured patronage plums in the Department of the Interior, and the real work of managing them did not begin until President Theodore Roosevelt, backed strongly by conservationists, fought and won a political battle to transfer the reserves to the Department of Agriculture's fledgling Division of Forestry, which was vigorously headed by Gifford Pinchot. Pinchot immediately proceeded to turn out the incompetents who had gained control of the reserves and to replace them with men of proven, practical ability. He declared that the purpose of his agency was to grow trees for lumber, not just scenery, and he sought to change the preservationist image of the reserves by giving them the more utilitarian-sounding name of National Forests.

Throughout the early years of the Pecos River Reserve and the Pecos National Forest, the main job of the government caretakers, or rangers, as they called themselves, was to control forest fires and to halt grazing and timber trespass. The job involved continual, sometimes dangerous confrontations with people who opposed the existence of the Reserves and Forests, and to be effective, a ranger had to have as much grit as he had intelligence and industry. He had to possess the self-reliance of the "bully" outdoorsman whom President Roosevelt so self-consciously praised, and no one who worked on the Pecos National Forest possessed these qualities in greater abundance than Tom Stewart. Stewart had grown up in the mountains on a ranch about halfway between Pecos and Cowles, and except that he lacked a squint, he even looked like Roosevelt. He was thickly and solidly built, with a wide, round head, and he wore a heavy moustache. His spirit of adventure had led him to join the Klondike gold rush of 1897, but like most other Alaskan fortune hunters, he returned home poorer than he had left.

Stewart returned from Alaska in time to serve James Wilhoit as packer and guide on his ill-starred reconnaissance of the high country, and a little later, in 1902, he joined the embryonic Forest Service while R. C. McClure was still supervisor. He started out as a ranger third class, and the bulk of his work involved fighting fires, building trails, and marking the largely unsurveyed boundaries of the Forest Reserve. He was paid

Photo 56. Crew of foresters in camp, 1907. Tom Stewart is on far right wearing badge and sixgun. Forest Supervisor Lee Williams at center. Several of the others are clearly not southwesterners; they might be recent graduates of Gifford Pinchot's new forestry school at Yale. Photo by L. Margolin. Courtesy Santa Fe National Forest.

sixty dollars a month for the months that were free enough of snow to let him work, and he was obliged to furnish all his own horses, food, tools and gear.

Gradually, as the scope of forest work broadened to include range and timber management, the local Hispanic villagers came to resent the rangers, sometimes fervently. They did not consider their traditional use of the mountains to be trespassing, as the rangers often did. Although at this time the Forest Service had not acquired large areas of patented Spanish or Mexican land grants, some unpatented claims had been included in the initial formation of the National Forests. To some extent, however, the issue of land grants was beside the point, for in the view of many villagers, all the National Forests and the land grant ejidos had essentially the same status. For generations these lands had constituted an important *de facto* commons that was generally considered to belong equally to all the people living in the mountains.

Proponents of the National Forests, of course, also believed that the mountain lands were a commons, but in their view its constituency included the entire nation. As rangers like Tom Stewart forcibly imposed this view on the locals, they came to be seen as agents of the same process of dispossession that men like Frank Bond and Alonzo McMillen were carrying out in land grant speculation. And indeed, the Forest Service would soon buy from the speculators many thousands of questionably obtained acres and incorporate them in the National Forests.

In protest, some villagers resorted to violence. Most rangers traveled in pairs, but even that was not always an adequate defense. Two rangers stationed in Rociada returned from patrol one day in 1901 to find that arsonists had set fire to their quarters. In 1909 two rangers in Cuba, both veterans of the recent campaign in the Philippines, were beaten so badly that their replacements were ordered never to go out of their homes without sidearms. And Stewart, for his part, found it prudent when dealing with grazing problems to camp two or three miles from the nearest herders and to sleep several hundred yards away from his campfire. That way, he figured, he would be aware of any late-night visitors before they were fully aware of him.

During his first years with the Forest Reserve, Stewart's efforts to control range trespass were frustrated by his inability to arrest or fine offenders. A ranger was legally empowered only to escort trespassers off the reserve, a task that sometimes took as long as two or three days. Then, once the herds were on private land and the ranger had taken leave of them, herders routinely turned their animals around and drove them back to the reserve. Stewart was greatly heartened, therefore, when word came that forest rangers had been granted the power to make arrests. He immediately purchased two pairs of handcuffs, a padlock, and ten feet of chain and started riding for the Rio Santa Barbara, where he expected to find the sheep of a certain local politician. As Stewart himself told it,

> Sure enough, I found three herders (two men and a boy) with sheep and no permit. They gave me the horselaugh, but I disclosed the joke was on them this time. I arrested the men and sent the boy home on a burro to get other herders.
>
> Camping for the night I handcuffed the prisoners to a tree. I felt certain their case would be dismissed in Santa Fe, so I decided they should get their justice on the way. After the new herders arrived and I gave the prisoners breakfast, I handcuffed each of the two, got on my horse, and marched them twenty-five miles on foot to Windsor's Ranch [Cowles]. From there we traveled by buckboard and train to Santa Fe. The case was dismissed, but it was soon rumored around that Rangers had police powers, and the trespass troubles became less numerous.

From 1909 to 1914 Stewart served as supervisor of the Pecos National Forest (which in 1915 was combined with Jemez National Forest and renamed the Santa Fe), and during that time he had the satisfaction of seeing considerable improvement in the ecological health of the high country. His hard-earned satisfaction, however, probably diminished considerably when the United States joined World War I. The order issued to the Forest Service to grant as many grazing permits as were requested wrecked the modest progress that had been made, and also vastly compounded the problems that postwar foresters inherited.

In the southern Sangres one of the men charged with the task of getting grazing back under control was John W. Johnson, the District Ranger at Pecos from 1919 to 1943. In 1920 and 1921, Johnson noted, "there were twice as many sheep allotted as should have been" on the upper Pecos. By then the Forest Service had begun to establish the grazing capacities of its ranges on more or less scientific grounds. A ranger calculated how many tons of forage a given pasture produced and what portion of it could be removed without reducing the vigor of the forage plants. Next he calculated how long it would take a given number of sheep or cattle to carry out the desired "harvest." There was naturally a great deal of guesswork involved in these determinations, and there was also a certain amount of fakery built into the procedures that underlay them. In order to justify the highest possible level of stocking, such plants as yellow pea (*Thermopsis divaricarpa*) and wild geranium (*Geranium caespitosum*), which sheep avoid except when starving, were classed as sheep feed. Johnson, who was more interested in reducing herds than in justifying their excessive numbers, chafed at having to engage in this kind of botanical equivocation, and the general public soon came to class him as antisheep, and eventually antihorse and anticow as well.

Not only were too many permits allotted for high-country

Photo 57. Gully at Las Lomas, near Santa Barbara Baldy in the Pecos High Country, headwaters of the Rio Pecos. 13 August 1926. Photo probably by Ranger J. W. Johnson. Courtesy Santa Fe National Forest.

Photo 58. View of Beatty's Park with stockman's cabin and corral, July 1924. Compare with Photo 24, p. 95, which was taken in 1977. Courtesy Santa Fe National Forest.

ranges, but many permittees ran more animals than they were entitled to. Johnson's problem, not an easy one, was to figure out how many livestock there actually were in the scattered meadows, forests, and canyons of the mountains. His method for solving it, however, had the kind of elegant simplicity that even his detractors were obliged to appreciate. He began by roping a calf and tying it in the middle of a meadow. Then Johnson waited while the bawling of the calf and its mother drew every nearby cow into the open. After counting the cattle according to brand, Johnson untied the calf and moved on to another meadow to repeat the procedure. By the time he finished, he not uncommonly discovered that some permittees were running twice as many cows as they were allowed, and the permittees, when confronted by Johnson, sometimes showed a dogged determination to explain that they were not at fault.

One rancher, whose seventy cattle in the high country were double his allotted number, told him, "Mr. Johnson, two years ago someone stole thirty-five cattle from me and now they must have come back."

Horses presented an even more difficult problem than cattle or sheep. Until Johnson's day it was a common practice to turn out old, lame, or unused horses to the high country to fend for themselves. As a result, large herds of so-called wild horses became year-round residents of the mountain rangelands. In winter they were especially destructive to vegetation as they congregated in relatively confined areas on south-facing slopes and grazed the available forage down to and including the roots. Johnson supervised repeated efforts to round up the horses. All of the horses he brought in were offered at auction, but there were always a few that were so broken down that not even their owners would redeem them. These unfortunate rejects were eventually destroyed. Then as now, the elimination of feral horses from public ranges was almost universally unpopular. The letter-writing public responded strongly to the symbol of "wild" horses but apparently had little feeling for the grasses and soils the horses destroyed or the native wildlife they preempted. Johnson himself was regularly cursed and threatened for his role in eliminating the feral horses, but he persisted nonetheless. Thanks in part to his efforts, the number of domestic livestock in the high country was drastically reduced, and many mountain meadows that grew little but sneezeweed and false skunk cabbage in 1920 supported a healthy cover of perennial grasses when he retired in 1943.

The conflict between the Forest Service and Hispanic villagers over National Forest grazing still continues, and it has generated, as most controversies tend to do, a good deal of misinformation that passes for history. For example, a 1976 teacher-training text repeats one especially persistent fiction by saying "the grazing lands in the National Forests and other government-owned tracts have been primarily issued to corporate out-of-state cattle growers." Actually, although immigrant Anglos did displace many Hispanic stockmen on the public domain of the eastern plains and northwest New Mexico, they did not often obtain permits for mountain grazing lands, and there is no clear evidence that they benefited in their efforts from the connivance of the Forest Service. Frank Bond and Edward Sargent, who eventually controlled a large number of permits, did so by obtaining them from Hispanos who were deeply in debt or otherwise hard-pressed for cash. They achieved their success not because of the desires of the Forest Service, but in spite of them.

In fact, the Forest Service makes a substantial effort to ensure that large cattle producers do not lease permits from smaller operators. Matching shipping records provided by the New Mexico Livestock Board with its own records identifying the brands of cattle permitted to graze on the National Forests, the Forest Service confirms each year that the owner of a given permit was also the owner of the cattle that grazed under it. Any departure from this strict rule of ownership is grounds for revocation of the permit in question.

In order to prevent the control of large areas of rangeland by a few wealthy individuals, the Forest Service also limits the maximum size of the permits it issues. This in turn controls the maximum size of single-owner herds on the National Forests. On both the Carson and the Santa Fe the so-called upper limit is about four hundred animal units, one unit consisting of a cow and her calf. This figure is approximately half as large as the upper limit permitted on other National Forests in the Southwest. While a forest's upper limit can be waived to accommodate special circumstances, no such waiver has ever been granted on either of northern New Mexico's two National Forests.

Today fewer than 50 of the more than 800 permittees of the Carson and Santa Fe National Forests are Anglo, and all of them are residents of northern New Mexico. Before 1940, when the total number of grazing permittees was double what it is now, the percentage of Anglos was even smaller. From the inception of its grazing program, the Forest Service sought to regulate existing patterns of range use, not to rearrange those patterns for the benefit of a single group. Where there was conflict in the awarding of permits, the Forest Service's policy in New Mexico and elsewhere was to favor local, small operators over large ones. Unfortunately in northern New Mexico the vast majority of ranching operations were small ones, and the ranges were so overstocked that the burden of herd reductions fell heavily on the already impoverished mountain villagers.

From the 1920s onward, and particularly from the 1940s to the 1960s, livestock ranching on the National Forests of northern New Mexico underwent a thorough transformation, and no trend through that period better illustrates the forces at work than the gradual abandonment of sheep and goat ranching by mountain villagers in favor of the raising of cattle.

As in every other sphere of economic life, the transition from a subsistence to a cash economy was a major factor in causing the change. On open rangelands sheep and goats require continuous herding, which makes raising them a very labor-intensive undertaking. So long as the animals are destined for household or local consumption and all other needs of the household can be met without cash, the large inputs of labor may not impose undue hardship. As soon as a family begins to raise livestock for the purpose of earning money, however, the calculus of benefits changes radically.

In seeking to enter the cash economy, the mountain villagers discovered that they, like stockmen throughout the West, were at the mercy of the changing tastes and needs of the national markets they served. They were compelled to learn that the prices they got for their animals did not keep pace with rising labor costs. This was the case whether they paid cash wages to others for assistance or they and their children provided all the necessary labor themselves. Either way, they had to be conscious that the time and energy involved in tending sheep might otherwise be devoted to the earning of cash wages through other lines of work. Over time, virtually every sheep permittee reached an unavoidable conclusion: cattle produced more profit for less work. However regretfully, the people of northern New Mexico gradually traded their flocks of skittish sheep for herds of cattle.

Table 1
Santa Fe National Forest
Grazing Permits and Livestock Numbers

	1940	1950	1960	1970	1980
Paid Permits	640	483	474	418	408
Free-use Permits	217	188	24	7	0
Cattle	7129	7580	8429	12173	11692
Sheep	27180	9532	4905	700	0

Table 2
Carson National Forest
Grazing Permits and Livestock Numbers

	1940	1950	1960	1970	1980
Paid Permits	921	897	696	490	427
Free-use Permits	461	239	13	78	0
Cattle	11497	9750	9580	10460	11972
Sheep	60300	46981	38292	26536	20637

Source: Office of Range Management, USDA—Forest Service, Southwestern Region, September 1984.

Another exceptionally important trend through this period involved the steady decline, alluded to above, in the total number of National Forest permittees. Whereas more than 2,200 individuals held permits to graze their animals on either the Carson or the Santa Fe National Forests in 1940, fewer than 1,000 still held permits in 1970. Once again outside forces were at work, for the changing economics of livestock ranching made the keeping of small herds of cattle or sheep less and less profitable. Profits were still further diminished by the severe herd reductions, as ecologically necessary as they may have been, which the Forest Service imposed on the ranchers. To make matters worse, these new measures were often excessively harsh, and the rangers failed to explain them in terms comprehensible to the villagers.

A second, less controversial measure adopted by the Forest Service also contributed to the steady decline in the number of grazing permittees. It required that when any permit for twenty-two or fewer cattle was transferred, the new owner had to be an existing permittee. The primary purpose of this rule was to prevent "outsiders" from acquiring permits, and thus to maintain the tradition by which ranchers from a given community enjoyed the exclusive use of the rangelands surrounding them. But a major effect of the transfer rule was to speed the consolidation of permits into progressively larger units, and to inhibit the perpetuation of permits for fewer than twenty-two head. Although from an economic point of view these small permits were unprofitable, they were, nonetheless, valuable to their owners, who surrendered them only with great reluctance.

However benign the intentions of the Forest Service may have been, its actions appeared malicious to the villagers and not uncommonly imposed severe hardship upon them. Nevertheless, the agency's primary objective continued to be restoration of the ecological health of the land, and the greatest obstacle it faced in this effort was that in spite of overwhelming evidence, most villagers refused (and many still refuse) to believe that the health of the land was at stake.

Over the years an awesome evolution has taken place in the Forest Service. In 1905 Chief Forester Pinchot issued a leatherbound book to his rangers containing the sum total of all the regulations and guidelines they were to follow. Called the Buckskin Manual, the slim volume was less than 150 pages long. Today's district rangers, however, have considerably more in the way of instruction to contend with as they manage the 100,000 to 150,000 acres in their care. Each of them has a bookcase filled with three tiers of thick binder notebooks that cover the broad outline of Forest Service policy, and in addition, one or two walls of file cabinets bulging with instructions for immediate planning and operations. No less burdened are the agency's Forest Supervisors, each of whom has jurisdiction over a National Forest comprised (in the case of both the Carson and the Santa Fe) of seven or more districts. And supervising the supervisors is the Regional Forester in Albuquerque whose area of responsibility includes all of New Mexico and Arizona and who reports directly to the bureau's highest level of management in Washington. As Congress willed it, the Forest Service has become a bureaucratic leviathan.

An unending proliferation of directives, analyses, and reports has obligated the men of the Forest Service (there are still few women in positions of responsibility) to devote as much of their time to paperwork as to fieldwork. Bearing little resemblance to their crusty predecessors, most Forest Service personnel today are highly trained technicians and managers, of whom an increasing number work behind desks at various forest and regional headquarters. Unlike the district personnel, who live on or near actual National Forest lands, many of these people see less of the actual land in their care than does the average tourist.

Yet for all the change it has experienced, the Forest Service is still distinctly different from other large and complex land-management agencies like the National Park Service and the Bureau of Land Management. From the outset, Gifford Pinchot intended to instill in his Forest Service an *esprit de corps* that would guarantee high standards of professionalism and relative immunity from political pressure. Toward that end he ordered that upper-level vacancies in the Forest Service be filled through promotion from the lower ranks rather than by recruitment of outsiders. Pinchot's policy has remained fundamentally unchanged for three-quarters of a century, and it has been famously successful. The Forest Service's reputation for integrity has discouraged political meddling, and morale within the agency has generally been high since talented young people advance in their careers with relative rapidity. But there has also been a negative side to the insularity of the Forest

Service. By closing itself off, the agency became unnecessarily inbred and resistant to change, and its institutional narrowness has in recent decades tarnished the agency's image.

World War II was a major watershed for the U.S. Forest Service. The agency's prewar reputation, built on solid achievements in conservation and preservation, slowly gave way to images of vast clearcuts, silted streams, and aerial spraying of herbicides and pesticides. Demand for timber skyrocketed after the war, and the wood products industry turned to the National Forests for an increasing share of the logs it needed. The major timber corporations had already cut most of their forests, and the second generation trees they had planted were not yet ready for harvest. It was time, their lobbyists argued, for the National Forests, which comprised 18 percent of the nation's commercial timber lands, to take up the slack. The Forest Service—and Congress—agreed. The agency's staff and appropriations increased dramatically as it undertook to build roads and arrange timber sales at an unprecedented rate.

Preservationists and wildlife groups were generally horrified by the new trend and claimed that the Forest Service had sold out to corporate interests and had become, in effect, a giant federal timber company. Actually, however, the Forest Service had not changed its institutional direction at all. Pinchot himself had written: "The object of our forest policy is not to preserve the forests because they are beautiful . . . or because they are refuges for wild creatures of the wilderness . . . but . . . the making of prosperous homes. . . . Every other consideration comes as secondary." The apparent change in the Forest Service was really only a change in the economics of supply and demand. Where before industry and preservationists had not competed intensely for use and control of the National Forests, now they did. And the Forest Service, because of its fundamental and sometimes single-minded commitment to tree farming and timber production, earned the enmity of the growing segment of the population that placed high value on the so-called amenities of wilderness, wildlife, and ecological integrity.

The conflict that resulted produced some of the most important environmental legislation in the nation's history: the Multiple Use–Sustained Yield Act of 1960, which required that all renewable forest resources be managed so that their productivity not decline in future generations; the Wilderness Act of 1964, which established a system of national Wilderness Areas to be protected from resource development; and the National Forest Management Act of 1978, which required detailed interdisciplinary planning for all management activities on the National Forests and which also provided for a high degree of public participation in the planning process.

The years of legislative achievement also featured numerous heated confrontations between the Forest Service and preservationists concerning specific land-use projects in National Forests across the country. One of the liveliest of these battles took place in the late 1960s and early 1970s in the southern Sangres at the edge of the Pecos Wilderness.

The controversy centered on the Santa Fe National Forest's plan to build a paved, two-lane highway from Terrero in the Pecos Canyon up across the high eastern divide of the range and then down the other side to connect with existing roads that led to Las Vegas. The initial justification of the road was its character as a "Scenic Highway," and scenic it would have been. From its highest point near the summit of 11,661-foot Elk Mountain, the road would have afforded car-bound tourists a view of nearly everything in northern New Mexico—a view that had hitherto been available only to backcountry hikers, horsemen, and occasional daredevil jeep drivers. But scenery alone could not justify the expenditure of the nearly five million dollars needed to build the road. For the final argument in favor of the road, the project's backers, including principally the Forest Service and a few declining War on Poverty agencies, turned to the idea of economic and social progress. The Elk Mountain Road, they said, would make possible the construction of a first-class ski resort atop Elk Mountain, which in turn would create more than seven hundred new jobs for economically depressed San Miguel County.

It was a heady promise, and as will be seen, a false one. The Forest Service became involved—and embroiled—in it primarily in order to obtain access to the sixty million board feet of harvestable virgin timber that grew on three sides of Elk Mountain. On the fourth and northern side of the mountain lay the protected forests of the Pecos Wilderness, from which Elk Mountain had been excluded because a rough jeep trail and a pair of abandoned mica mines (including George Beatty's) scarred its summit. In every other respect, as opponents of the road earnestly pointed out, Elk Mountain was just as wild and pristine as the Wilderness. A lot of people thought it should stay that way.

Preservation, however, was not the only or even the most compelling issue in the fight over the Elk Mountain Road. Opponents of the project, who joined together to form the Upper Pecos Association, were alarmed at the lack of clarity in Forest Service planning that allowed the commercial attractions of sixty million board feet of timber to justify, and at the same time be justified by, a multi-million-dollar ski resort. The controversy became an environmental equivalent of the classic riddle of the chicken and the egg, as no one was quite sure which of the two projects had given birth to the other. Even more disturbing, however, was the question of whether the planned ski resort, which had failed to attract the serious interest of developers, would actually produce the jobs that were claimed for it.

It turned out that it would not. In order to acquire the 3,000-foot vertical drop necessary to compete with other New Mexico ski resorts, the Elk Mountain Ski Area would have had to borrow land from the Pecos Wilderness, which Congress, as the ultimate caretaker of the Wilderness, would hardly allow it to do. As a result, any ski facility on Elk Mountain would have amounted to little more than an extravagantly expensive beginners' hill, and the difficulty of relating seven hundred or more jobs to so paltry an installation drove project backers into the netherworld of advanced mathematics. San Miguel County, however, like all the rest of northern New Mexico, did need jobs—and desperately. Unemployment in the county in 1967 hung at 11.7 percent, and nearly 40 percent of San Miguel's families lived in what the government officially defined as poverty. But the Elk Mountain Road, on close inspection, seemed unlikely to be much help. At the cost of four million dollars for the road, several millions more for the ski area, and a host of environmental problems including soil erosion, loss of habitat for the Pecos elk herd, and overcrowding of the Wilderness itself, the road seemed likely to create only a handful of new jobs, many of which would be taken by newcomers to the area, not by residents. The only clear beneficiaries of the road would be the timber merchants of the Forest Service and the local logging companies.

Fortunately for Elk Mountain, the local residents, and the taxpayers of the United States, the Upper Pecos Association was ultimately successful in its efforts to stop the Elk Mountain Road. At the end of nearly a decade of wrangling over the project, the Forest Service had nothing to show for its efforts except a black reputation as an enemy of wilderness in northern New Mexico. And the Hispanos of the Pecos and Las Vegas areas, whom everyone on both sides of the issue had claimed to want to help, were no better off than before. What seems peculiar about the controversy of the Elk Mountain Road is that, after the project was defeated, no one came forward with alternative plans for helping the poverty-stricken people of San Miguel County. If the government was ready to spend four or five million dollars for an ill-conceived road, why was it not also willing to spend that money in a genuinely useful way? The Forest Service, for the time being, had no answer.

16

Land and Cattle

*Most of all I beseech you to sense the urgency of
the situation in northern New Mexico.*
> —William Hurst, Regional
> Forester, 1969

TOURISTS IN THE SOUTHERN SANGRES know that they have finally *arrived* in the mountains when they pass one of the
handsome wood and stone signs that say: "Entering Carson
[or Santa Fe] National Forest—Land of Many Uses." The sub-
title strikes just the right tone. It makes the green mountain
world seem like a coniferous Oz: full of variety and wonder
and better than other places. If the weather holds during the
visitors' stay and no one vandalizes their car (Texas plates run
an extra risk), they will probably leave the mountains feeling
that their first impressions were accurate. The National Forests truly deserve their reputation as great vacationlands.

To the Hispanic people who live in and around the National
Forests, however, those roadside proclamations convey a more
complex message. For them the Land of Many Uses is a geographic and political entity in many ways more powerful than
their county. The Santa Fe and Carson National Forests together comprise more than two and a half million acres of
northern New Mexico, including 44 percent of Taos County.
With a shrug of its immense bureaucratic shoulders the Forest

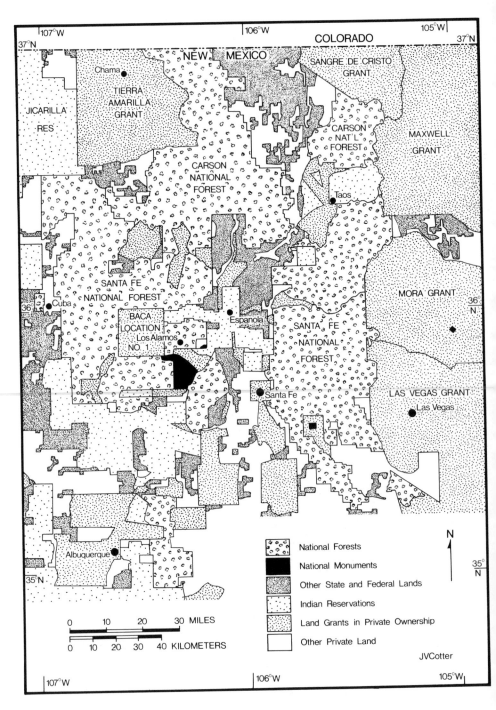

Service can create or destroy, move or let stay nearly everything that matters: jobs, roads, livestock, even landscapes.
Moreover, many Hispanos feel that the Land of Many Uses is
a counterfeit kingdom. They say that much of the land claimed
by National Forests belongs, in fact, to them.

No one can dispute that hundreds of thousands of acres once
did. Twenty-two percent of the Carson and Santa Fe National
Forests consists of patented Spanish and Mexican land grants,
and a large additional acreage includes grants that Hispanos
claimed but U.S. courts never confirmed. The unconfirmed
claims were incorporated into the National Forest System along
with other portions of the public domain, while the patented
grants were obtained through outright purchase. Contrary to
popular belief, the Forest Service did not acquire these grants
directly from the villagers living on them. Rather, it was usually second or third in the chain of ownership following the
original sale of a grant, often having obtained the grant from
a timber company, which in turn had bought it from a speculator. Nevertheless, as the ultimate owner of many land grants,
the Forest Service inherited much of the discord and bitterness
that the genuinely sordid business of land grant speculation
generated. The Forest Service may not have been a perpetrator
of land grant crime, but in many people's eyes it was an ac-
cessory after the fact. It was also large, public, and permanent—qualities that most land grant speculators were happy
to have lacked.

No other National Forests in the United States are plagued
with as unhappy a historical legacy as those of northern New
Mexico, and none exist in as complex a cultural environment.
None can claim, as these can, that because of traditions that
are centuries old, the cutting of firewood in the piñon and

Map 7. Land ownership in north central New Mexico.

juniper woodlands generates more public controversy than the management of commercial sawtimber. And certainly no forests but these can claim to have had a Reies Tijerina.

Reies López Tijerina is the charismatic founder of the Alianza Federal de las Mercedes, a union of Hispanic New Mexicans whose goal for twenty years has been the restoration of the old land grants, the *mercedes*, to the descendants of the original owners. A key element of Alianza doctrine is that the grants are more than mere tracts of land; they are independent social and political entities. Accordingly, Tijerina and his followers translate the name of their organization as "the Federal Alliance of Free City States."

Born in Texas in 1923, Tijerina spent much of his early adult life as an evangelist. By his own account, he fell afoul of the Assembly of God because he preached that the church should give to the poor and not the other way around. Tijerina settled for a time in Tierra Amarilla in northern New Mexico, but restlessness and poverty caused him to move on, and for several years he and his family wandered across the West and Middle West as migrant laborers. Tijerina's nomadic life culminated in a six-year sojourn in Mexico, where he spent much of his time studying land grant law. In 1963 he returned to New Mexico and organized the Alianza.

Tijerina and the *valientes*, or militants, who joined him, borrowed heavily from the tactics of civil disobedience that black civil rights activists were then using with great effectiveness in the South, and they staged numerous protests and marches while calling for the readjudication of land grant claims. Along with the issue of land ownership, the Alianza also voiced grievances concerning the management of grazing on the National Forests and especially the discontinuance of the so-called free use permits.

The free use permit was one of two broad classes of permit that the Forest Service traditionally administered. The standard permit, for which the permittee paid a fee, applied to the grazing of livestock for essentially commercial reasons: the production of meat or fiber. The Forest Service issued free use permits, on the other hand, for animals, such as plow horses and milk cows, that were essential to the day-to-day operation of local households. Free use permits helped to perpetuate the traditional subsistence way of life in the villages, but their environmental cost was extremely high. Milk cows and draft horses necessarily grazed close to the villages, and as a result the land surrounding many settlements was virtually barren. Many of these animals, moreover, were "retired" cows that gave no milk or horses that no longer worked, and in the view of the Forest Service, keeping these animals constituted an abuse of the free permit system. In the 1960s, judging that such abuse was epidemic and that the economic decline of subsistence farming had destroyed the only remaining rationale for free use permits, the Forest Service gradually phased them out. The villagers, naturally enough, became enraged. Many of those animals still served an economic function, and even if they did not, ownership of them was symbolically important. And sentiment was also involved, for some of those cows and horses, in spite of their broken-down and half-starved condition, were pets.

The issue of grazing management provoked even more outrage as the Santa Fe and Carson National Forests imposed severe reductions of standard permits on many allotments. The people of Cundiyo, whose cattle grazed the Sierra Mosca allotment on the east side of the Pecos Wilderness, saw their herds cut by 60 percent. Their counterparts in Canjilon, a tiny and remote village between Abiquiu and Tierra Amarilla, fared even worse, as they lost permits for a thousand cattle in the course of just a few years. The Forest Service justified these and many other stringent reductions on grounds that without them the overgrazed ranges would eventually lose their capacity to support herds of any size. Forest Service studies showed that although 21,600 cattle and 32,200 sheep were permitted to graze on the Carson and Santa Fe forests at that time, the range lands were capable of supporting only 14,400 and 25,200 respectively. The villagers, however, were deaf to the rangers' arguments about protecting the rangelands. Their own economic health, never very robust, was under attack. To most of them the reductions seemed blatant acts of political and social aggression.

No one had to tell the villagers that the grazing issue and the land grant issue were essentially the same thing. They knew that their ancestors had grazed their herds and flocks without restriction, and most of them believed that if they could get their land back from the Forest Service, their worst problems would be solved.

The land claim on which Tijerina and the Alianza ultimately focused their greatest energies concerned the San Joachín del Rio de Chama land grant, located north of the village of Coyote in Rio Arriba County. The Alianza argued that the Court of Private Land Claims should have confirmed the grant for 500,000 acres when it ruled on the claim in 1896 and that the 1,423 acres that the court actually did confirm were unlawfully alienated from their legitimate Hispanic owners. Alianza protesters staged "camp-ins" on the disputed land in 1965 and 1966, and on the second occasion the Forest Service elected to force the issue of ownership by attempting to collect camping fees from them. Tijerina and his people retaliated by "arresting" two forest rangers for trespassing. In short order the state police arrested Tijerina and several others on charges of assault and battery and expropriation of government property. From that point on, the politics of confrontation in Rio Arriba County escalated rapidly.

A balanced and well-informed history of this turbulent period has yet to be written. Virtually all published accounts of Alianza activism during the late sixties have been highly journalistic. They have tended to place excessive emphasis on the personal activities and pronouncements of Reies Tijerina while neglecting to assess other leaders and the diversity of opinion and ambition that existed within the group. This oversight, however, is understandable considering that many Alianza activists have been prudently reticent about their controversial and often illegal exploits and considering also that Tijerina himself possessed an impressive natural ability to capture—and to manipulate—the attention of the media.

It will take a book at least as long as this one to tell the whole story of the Alianza movement, which enjoyed a far broader base of popular support than most writers have attributed to it. The Alianza movement is important not just because of the dramatic events it produced in 1966 and 1967, but because those events in turn stimulated real change in the way the Forest Service and other agencies treated the Hispanic villagers.

Through the fall of 1966 and the following spring, arsonists repeatedly set fire to the homes, barns, and hayracks of Anglo ranchers in Rio Arriba, some of whom, fearing bloodshed, moved out of the troubled county. Amid talk of "direct action," Alianza leaders scheduled a mass meeting for June 3, 1967, in Coyote. County District Attorney Alfonso Sánchez feared that insurrection was imminent and forbade the Alianza to assemble. He further moved to break the group's strength by ordering the arrest of Alianza leaders on the night before the scheduled meeting. When lawmen subsequently hailed into court more than a dozen members of the group, not including Tijerina, the stage was set for the most serious civil violence in New Mexico's recent history.

On June 7 a band of twenty armed men entered the sleepy, decaying town of Tierra Amarilla and stormed the courthouse. They aimed to make a citizen's arrest of District Attorney Sánchez, on grounds that he had abused the constitutional rights of the arrested Alianza members. Fortunately for Sánchez, he was not in the building at the time of the raid, but in the course of their search for him the raiders shot and seriously wounded a policeman and a jailer. Two other policemen were beaten, one seriously. The county judge, meanwhile, escaped mayhem by locking himself in a bathroom.

What followed was the largest manhunt in New Mexico's history. On the order of Governor David Cargo, National Guard tanks and artillery rumbled into Rio Arriba County, supported by 200 military vehicles, 350 soldiers, and scores of police, cattle-brand inspectors, and other armed lawmen. The invading army converged on the tiny village of Canjilon, long a stronghold of Alianza support. But instead of Tijerina and the others who had participated in the courthouse raid, they found an unarmed and peaceful band of Hispanic picnickers whom they detained, in spite of rain, in a corral. Resentment of the government's heavy-handed tactics grew as the search widened. In Santa Fe, meanwhile, the behavior of some political and law enforcement officials verged on hysteria as rumors flew wildly about to the effect that the Alianza's activities were masterminded by a cadre of Cuban infiltrators and that the rebels' next move would reveal their possession of enormous caches of arms. Every day was a frenzy of activity, but still the fugitive raiders remained free.

Not until two weeks later, after receiving an anonymous tip, did police arrest Tijerina in Albuquerque. He was booked on two counts of assault to commit murder, two counts of kidnapping, one count of destruction of state property, and

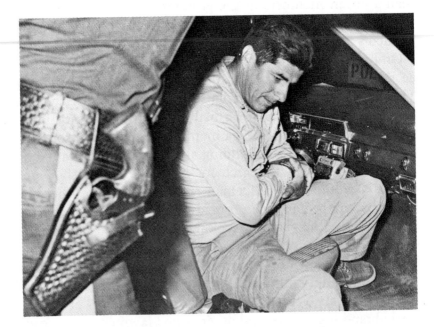

Photo 59. Reies López Tijerina. Photo by Bud Jorgenson.

one count of possession of a deadly weapon. The nineteen others who had been with him at the Tierra Amarilla courthouse were also eventually arrested and charged with similar crimes.

Sadly, the violence was not over. At a preliminary hearing, jailer Eulogio Salazar, who had been shot in the face during the raid and permanently disfigured, was asked who had wounded him. Salazar identified Reies Tijerina. Before the case came to trial, however, and while all the "Tierra Amarilla twenty" were free on bond, Salazar was bludgeoned to death and his body abandoned in his bloodstained car. His murder was never solved. Nearly a year after his death, in December 1968, an Albuquerque jury surprised the rest of the state by finding Tijerina innocent of all charges stemming from the courthouse raid.

As one would expect, the raid provoked a flurry of investigations. Undersecretary of Agriculture John Schnittker flew to New Mexico to have a firsthand look at Forest Service operations—no doubt to prepare the USDA's response to inquiries from other branches of government. A few weeks later, in August 1967, Senator Joseph Montoya of New Mexico opened hearings of his Special Subcommittee of the Senate Committee on Public Works in Albuquerque, Taos, Española, and Las Vegas, and many of the witnesses who appeared before the subcommittee registered grievances about Forest Service policies. A state representative from Taos County, for instance, claiming to speak for ranchers in his district, complained that the Forest Service refused to make the abundant forage resources of the National Forests fully available to local stockmen. If the editorial cartoon that appeared in the Bernalillo *Times* a week after the conclusion of the hearings is any indication, the Representative's views were widely shared in northern New Mexico.

Finally, in October 1967, Secretary of Agriculture Orville Freeman came to El Paso to attend a conference on Mexican-American affairs and had occasion to listen to still more enraged protests about the allegedly insensitive land manage-

Figure 1. From the Bernalillo Times *(Sandoval County), 27 August 1967. Significantly, the cartoon does not suggest why the rancher's land is so poor.*

ment practices of the Forest Service. As a result of his experience at the conference, Secretary Freeman instructed the Forest Service to

make an even stronger effort to work in rural development and poverty programs to help the Mexican-American people. There is pathetic poverty, great ignorance, unfortunately emotional unreasonableness too, but the Forest Service, I am sure, is big enough and tolerant enough to live with that and to help people who need help.

Freeman further asked the chief of the Forest Service to involve himself personally in the development of grazing and recreation budget requests for northern New Mexico. Freeman was saying, in effect, that the Department of Agriculture was prepared to increase spending considerably in order to ease the poverty—and quiet the civil unrest—of the rural Hispanic villages. During the next fiscal year the subsequent doubling of budget allocations for range improvements in the Santa Fe National Forest to $235,000 proved the depth of the department's new commitment.

As a further result of Freeman's interest in northern New Mexico, Regional Forester William Hurst directed a member of his range management staff, Jean Hassell, to evaluate how the National Forests might better serve the needs of Hispanic villagers. Hassell's report, "The People of Northern New Mexico and the National Forests," was completed four and a half months later and immediately became the basis for new National Forest policies in northern New Mexico.

Hurst had made a good choice. Jean Hassell by then already possessed a reputation inside and outside the Forest Service for both firmness and fairness. Big physically, ranch-raised, and familiar with the land and the people of northern New Mexico, he proved to be an able critic of his organization. The evidence also indicates that he did so without provoking undue resentment, for a year later he became Supervisor of the Carson National Forest, and in 1972 he succeeded Hurst as Regional Forester.

Recognizing that the people of northern New Mexico were "behind the rest of the country socially and economically," Hassell predicted that the Hispanos' problems might be solved through "entrance . . . into the American mainstream of life," but he did not advise that entrance be made at the cost of the Hispanos' cultural integrity. Hassell argued that the Forest Service, as the largest landowner in the region, had an obligation to contribute to the economic advancement of the Hispanos, and he specifically recommended nearly 100 measures, 25 of them relating directly to grazing, that the Forest Service should undertake toward that end. Hassell urged that the Forest Service continue the high level of funding for range rehabilitation and improvement that had been initiated soon after the courthouse raid, but he warned against the kind of crash program that in recent months had already led to the construction of a number of stock tanks and pasture division fences in unusable or inappropriate locations. He also advocated that the Forest Service abandon its policy of forbidding the transfer of small permits (twenty-two animal units and less) to nonpermittees on grounds that the restriction obligated permittees to sell their permits on an artificially limited market where they could not receive a fair price. No less important, elimination of the transfer rule would help to perpetuate the existence of small permits and might help slow the decline in the total number of permittees on the northern Forests. Finally, Hassell also urged that where possible, the Forest Service split up large permits into thirty-cow units, which it could then distribute to the poor.

The main thrust of Hassell's recommendations was to call for greater sensitivity and flexibility in the way the Forest Service carried out its work, and one of the Forest Service practices that he singled out for criticism indicates how petty and bureaucratic the agency could be. Incredibly, prior to the Hassell Report, the Forest Service had been billing villagers for trespass fines as low as twenty cents, a procedure as wasteful of administrative energy as it was destructive of the public's goodwill.

Regional Forester Hurst endorsed the Hassell Report and ordered the immediate implementation of sixteen of the recommendations, including the lifting of restrictions on the transfer of small permits. He then assigned the rest to "further study" (where many of them, notably the proposal to divide large permits, have remained). The incomplete implementation notwithstanding, Hurst made his support of the new direction in forest policy abundantly clear, and he ultimately expanded its argument for cultural preservation to include Indian cultures as well. In 1972 Hurst summed up the Forest Service's new role as cultural conservator by saying,

The unique value of Spanish American and Indian cultures in the Southwest must be recognized and efforts of the Forest Service must be directed toward their preservation. These cultures should be considered "resources" in much the same sense that Wilderness is considered a resource with Forest Service programs and plans made compatible with their future well-being and continuance. . . . Forest Service employees at all levels of the organization must have a burning desire to perpetuate these unique values.

Whether their desire burned or only smoldered, the men and women of the Forest Service eventually altered two important trends in the management of Carson and Santa Fe National Forest rangelands. First, the rate of decline in the number of grazing permittees slowed; second, the Forests' total grazing obligation—that is, the amount of forage it committed itself to provide to permittees—ended its long-term steady decline and began to increase.

The Forest Service was able to increase its obligation to its permittees primarily because it had managed, thanks to enormous investments of money and manpower, to increase the capacity of its rangelands to provide forage. In some cases it drilled wells and built storage tanks and pipelines that brought water to previously unusable grazing areas. In others, it divided allotments into two or more pastures so that each could be rested from grazing periodically in order to coax maximum productivity from the range plants growing there. The water improvements and the rest and rotation systems of pasture management, however, required that the habits of animals and their owners be changed. Cows had to be forced from their *querencias*, the favored places where they preferred to graze, and their owners had to be persuaded to herd them more actively, to lure them into underutilized areas by moving salt blocks from place to place, and to build and maintain many miles of new fence.

Not surprisingly, Forest Service rangers and range conservationists found a lot of permittees unwilling to accept the new management systems. The moving of salt and cattle from one place to another required significant additional inputs of labor, and many permittees argued that the Forest Service already demanded enough of them in the way of grazing fees and general maintenance.

Another means by which the Forest Service managed to increase its capacity to provide forage was even more direct:

it created new rangelands. Small-scale range conversion work began in the early sixties and initially involved natural meadows where shrubs or trees had invaded as a result of overgrazing. With the advent of the Le Tourneau Tree Crusher and other land-clearing technologies in the late sixties, however, range conversion projects became much more extensive—and expensive. On both the Carson and Santa Fe National Forests large areas that supported piñon-juniper or sagebrush ecosystems were physically stripped of their vegetation and then sown to forage plants. On Rowe Mesa, directly south of Pecos, for instance, the Forest Service cleared thirteen thousand acres of woodland and reseeded them with grasses. A substantial portion of this new capacity now accommodates cattle that formerly grazed in the Pecos Wilderness. By creating the new grasslands, the Forest Service essentially eliminated overgrazing in the high country and eased conflicts between ranchers and recreationists.

As one would expect, these efforts were warmly received by the permittees who directly benefited from them, but other villagers were not so content with the changes. They recognized that for the sake of a relatively few permittees, the Forest Service was sacrificing large areas of piñon-juniper woodland that consituted a fuelwood resource of benefit to the entire community. The oil shortages and rapidly rising energy costs of the 1970s sharpened this realization further and eventually the Forest Service, too, lost its enthusiasm for the wholesale manufacture of new grasslands. Since then, in fact, a number of Forest Service range managers have begun to look back with regret to some (although certainly not all) of the range conversion projects they supervised. Like many of the permittees they work with, they are aware that their excesses, no less than their triumphs, are in part a legacy of the Courthouse Raid.

It is no secret that range improvements are costly. Between 1971 and 1980, the Santa Fe National Forest spent more than $2.5 million on range improvements. More significantly, few, if any, of the projects undertaken were expected to be cost effective over their useful life in any strict economic sense.

That is to say, when benefits are calculated solely in terms of dollars, most of these improvements will return only seventy or eighty cents for every dollar invested in them. The reason behind these economic inefficiencies is that when evaluating a proposed improvement, Forest Service administrators take into account the social and environmental benefit that they expect the project to generate, and, as Jean Hassell urged in his report, they calculate those benefits more liberally with respect to northern New Mexico than anywhere else within the Southwest Region. In effect, the Hispanic ranchers of the Carson and Santa Fe National Forests, the vast majority of whom are extremely small-scale ranchers, receive an indirect but substantial federal subsidy for their livestock operations.

Interestingly, the ranchers themselves also subsidize their operations, for even taking into account the meat produced for home consumption, most of them raise too few animals to obtain a net positive return on the time and money they invest. Although fifty cows are usually considered the minimum necessary for a commercially viable operation, the average size of herds grazed on the Santa Fe National Forest in 1979 was only eighteen head. Small livestock ranchers, like those of the Hispanic villages, suffer proportionately more than commercial ranchers from rising fuel costs, grazing fees, and maintenance responsibilities, and from theft and natural hazards. Their investment of labor per animal is extremely high, yet because of poor breeding stock and other limitations, their calf crops, and hence their income per animal, tend to run well below average. Nonetheless, there are hundreds, if not thousands, of individuals in northern New Mexico who would gladly pay a high price to obtain a grazing permit for a few head of cattle on National Forest rangeland. Unfortunately for them, permits seldom change hands, and the limitations of the land prevent any substantial increase in the number of permits issued.

As unprofitable as the average ranch may be, a very few permittees with herds as small as thirty do manage to support themselves from the proceeds of their livestock operations. These remarkable individuals tend to be astonishingly frugal and to prefer the simple, cash-poor life-style of their grand-parents to the restlessly consumer-oriented conventions of the present. The vast majority of permittees, however, work full-time jobs during the week and are only weekend ranchers, conceiving of their small herds as a kind of savings account or emergency fund on the hoof. They persist in ranching not because of its economics but in spite of them. What they gain is the pleasure of outdoor work and contact with nature and the opportunity to keep alive an ancient tradition of ranching on ancestral grounds. Among the other, no less important, reasons for being a permittee is the satisfaction of being "free of the Safeway," at least as far as meat is concerned. Another is the social status of being a rancher, which includes being

Photo 60. Driving cattle on the Sparks Trail to summer range in the Pecos Wilderness. Few small-scale ranchers in northern New Mexico consider money to be the primary reward of ranching. Author photo.

able to drive down the main road of the village with saddled horses in the back of the pickup and thereby to demonstrate that one is a man of the land and of property—no small thing in a region where welfare is a leading source of income.

By any criterion, the Forest Service has made great progress in restoring the ecological health of its northern New Mexico rangelands. In the late 1950s the Santa Fe National Forest was overstocked by nearly 100 percent, which is to say that the number of cows that grazed on it was double the number it could biologically support without losing productivity. By 1980, however, overstocking on the Santa Fe had declined to less than 5 percent, and the figures for the Carson Forest were similar.

Unfortunately, this dramatic progress has been won at considerable cost to Hispanic New Mexicans. Lost for many is the economic security of self-sufficiently producing meat for the family table. Lost or reduced is the sense of being connected in a fundamental, life-giving way to the land and to the traditions of one's ancestors. And lost or reduced is the pride that comes from feeling that the land substantially remains a collective Hispanic possession.

The bitterness engendered by these losses continues to sour relations between villagers and the Forest Service, and in a paradoxical way, makes the likelihood of more such losses greater. Although grazing on a forest-wide basis has declined to a more or less acceptable level, some individual allotments remain seriously overstocked, and most of these have little or no potential for realizing additional capacity through revegetation or the development of new sources of water. As a result, the only remaining strategy for increasing capacity, and thereby avoiding herd reductions, is to adopt improved management systems involving multiple pasture rotations and other labor-intensive procedures. Not surprisingly, many Forest Service range conservationists have been unable to convince their permittees of the merit of these grazing plans—or of anything else. Said one range specialist, who was somewhat renowned for the patience and skill with which he made his presentations,

CHAPTER 16

I could not, with the aid of calendars or all manner of dissertation or sworn witnesses, convince the permittees in that town that it was, in fact, Saturday afternoon. They knew, given their lifelong mistrust of bureaucrats in general and Forest Service people in particular, that if they agreed it was Saturday, they had made a fatal concession and the Forest Service would get to them *some way*. It was the most uncanny thing for me. I, as a person, had given them no reason to doubt my word, but they simply would not believe me, nor the ranger, the range staff in Taos, nor the supervisor himself.

In a sense, the permittees are right. To agree with the Forest Service, even on a small matter, is to concede yet another point in the continuing debate over who really owns the mountains and mesas. It poses the possibility that non-villagers know better than villagers how to manage the traditional grazing of livestock on the ancestral lands. Already the influence of Anglo-American society has destroyed so much of the autonomy of the village world that to accede to Anglo thinking in so traditional, almost sacred, an activity as ranching might threaten for some villagers the survival of all that remains. To make matters worse, Forest Service personnel change jobs frequently and move from district to district in order to advance their careers. As a result, the permittees are not always certain that agreements made with one ranger will be upheld by his successor, and in any event, they reject the idea that people who are transient members of their community should tell them what to do. As one permittee put it,

We are natives. We have dealt with grazing here all our lives. But the Forest Service works different. Rangers change all the time. And they always change the rules so we are always starting over. Each new bunch has different ideas. We know how the ball bounces. Every time a new guy comes in, we have to fight his ideas. He may have a college degree, but we know the country. I've known it since I was a boy, and my boy, when he is my age, will know it as well as me.

For seventy-five years the Forest Service and the mountain villagers have struggled to convince each other of the validity of their respective perceptions of the land and of society's proper relation to it. And also for seventy-five years the man-

Photo 61. Cerro Pedernal near Abiquiu. Hard use of the land undercuts the stability of northern New Mexico's rural communities. Author photo.

agement of grazing on the National Forests of the southern Sangres has posed a seemingly unresolvable paradox, which psychologists would call a "double bind." Protection of the land has meant injury to the village people, and protection of the villagers' cherished pastoral traditions has meant injury to the land. It has been impossible for any course of action to be kind to both land and people at once. Very few northern New Mexicans, regardless of ethnic group, have been willing to accept this simple though unpleasant truth.

Many well-meaning Hispanos and equally sincere Anglos continue to argue that for a variety of evil reasons the Forest Service willfully and needlessly destroyed the pastoral base of the village economy and substantially deprived the villagers of the use of their ancestral lands. In some cases this malicious behavior is attributed to the desire of the Forest Service to cater to the interests of large-scale Anglo ranchers. In others, the agency is alleged to be making more resources available to timber interests or to be reserving the mountains as a playground for urban recreationists. Like most easy answers, these ignore the fundamental complexity of the grazing problem.

In spite of its many blunders and its periodically stupid, oppressive, and imperious behavior, the Forest Service has not been an enemy of the people, and on the whole it makes an unconvincing scapegoat for the woes of the village world. One must ask what would have become of the mountain world had it never been "invaded" by the Forest Service or the collectivity that the Forest Service represents. Would it have been a land of stable, pastoral villages living in harmony with the surroundings?

Clearly not. The mountain villages were never particularly stable. Ever since the frontiers of the village world ceased expanding in the second half of the nineteenth century, its people have struggled to cope with demographic realities. With the frontiers fixed, surplus population could no longer "hive off" to form new villages, for every habitable scrap of land was already occupied. Necessarily the burden that the people placed upon the land increased. Their herds grew rapidly to keep pace with their needs, but except for new grasslands created by occasional forest fires, their resource base did not grow at all; in fact, with overuse, its productivity declined. And so arroyos formed, topsoil washed to the Rio Grande, and mountain streams grew wilder and more dangerous. If there had been no Forest Service, the process of deterioration in all likelihood would have advanced much further—as in fact it has on lands controlled by the Bureau of Land Management, the State of New Mexico, and a few remaining land grant associations, where range and timber management has been lax or nonexistent.

Those who argue that the Hispanos of northern New Mexico are essentially a Third World people who face the Third World problems of economic imperialism and cultural aggression should bear in mind that the most pressing and most serious problems of the Third World have little to do with imperialism or ideology. They are the demographic and environmental problems of desertification, deforestation, and erosion, all of which undermine the capacity of the land to meet the needs of growing populations.

For an example of how rapid and unpolitical the process of resource destruction can be, one need look no farther than the piñon and juniper woodlands of the southern Sangres. Valuable for centuries to Pueblo Indians and Hispanic villagers for both nuts and fuel, the woodlands were devastated in the late 1960s and 1970s by unrestrained firewood cutting as well as by the range conversion work mentioned above. Historically, fuel collection caused local deforestation only in the immediate vicinity of the settlements, but with the advent of the gasoline-powered chainsaw and the four-wheel-drive truck, the environmental impact of the woodcutter increased dramatically. Soon the growl of his saw was heard in every canyon and on every ridge. Since piñon and juniper trees held no interest for the timber industry, the Forest Service did not manage them for sustained yield but allowed villagers to harvest them as they pleased. Indeed, following the Alianza protests and the Courthouse Raid, the Forest Service sought to develop the woodlands as an economic resource by working with the federal Office of Economic Opportunity to establish wood cooperatives through which village woodcutters could market their harvests collectively for a good profit in urban centers like Albuquerque and Denver. The wood co-ops prospered for a time, but most developed serious management problems and ceased operating by the early seventies. Demand for green piñon and juniper wood, however, stayed high as a result of oil shortages and rapid increases in the cost of butane and propane gas, and the indiscriminate cutting continued.

Piñon is especially valued for firewood because, being slow growing, its wood is dense and resinous and burns with a long-lasting, even heat that makes it ideal for cooking. Unfortunately, since piñon trees grow so slowly—on the average, requiring 180 years to achieve a six-to-eight-inch diameter at the root collar—the woodlands could scarcely produce enough trees to satisfy the demand for wood.

Communities tended to become aware of this problem in direct proportion to the scarcity of fuelwood around them. One area where fuelwood became extremely scarce was among the lower, semi-arid slopes of the Sangres around Las Trampas, Chamisal, and Peñasco. In 1977, responding to pressure from local residents, the Peñasco District of the Carson National Forest inventoried its woodlands and discovered that about two hundred and fifty cords of green piñon and juniper wood could be harvested annually from the district on a sustained-yield basis. A check of the permits issued at the ranger station, however, showed that the Forest Service had authorized a harvest for that year of over 1,700 cords, and since many local residents typically neglected to obtain permits (although the permits were free, or nearly so), the actual harvest for that year was undoubtedly much higher. The figures indicated that the woodlands were being cut down much faster than anyone had anticipated. For all practical purposes, they were being mined. Furthermore, contrary to popular opinion, the wood-cutters did not include a large number of "outsiders" from Santa Fe and Albuquerque, or even Chimayo and Embudo. Almost all of the permits were signed by villagers who lived within the boundaries of the district.

Residents of other districts were also discovering that their woodlands were similarly depleted and that the harvest of green piñon and juniper trees would have to be curtailed or sharply reduced. In anger, many people criticized the Forest Service for having failed to restrict the harvest sooner and for having cleared large acreages of woodlands in order to create more forage for livestock owned by village ranchers. No one appreciated the irony of the new complaint more than the Forest Service range managers who had been excoriated only a few years earlier for imposing too many restrictions on ranchers and for not making more forage available to them. They had argued that without management there would be no range resource left to use. And now, after years of unrestrained harvesting, during which the Forest Service had failed to recognize fuelwood as a resource requiring management, the piñon and juniper woodlands were decimated, the fuelwood resource in many areas exhausted.

In the years since the raid on the Tierra Amarilla courthouse, the Alianza has split into several rival groups, but the issues that it protested have never died. At the heart of every argument over resources and every discussion of how best to conserve the integrity of village culture lies the question of who, by rights, owns the land. In 1979 Reies Tijerina was back in Coyote to call again for the readjudication of the San Joachín Land Grant. He and his supporters erected roadblocks and prevented logging trucks from entering the alleged lands of the grant. Again his efforts attracted headlines and caused alarm in the Forest Service, but again he failed to persuade the courts to take up the land grant issue. Neither he nor his compatriots, however, are likely to stop trying. As long as large numbers of rural Hispanos are poor and embittered, and as long as they have meager prospects for a better future, the issue of land tenure will flare again and again.

There seem to be two lessons that emerge from the story of land and cattle in northern New Mexico. The first is that in some instances a measure of ecological harmony and stability can only be won at the painful cost of cultural and social conflict. It is also clear, however, that a society cannot long preserve its culture without also conserving the resources that give it life. People everywhere, not just Hispanic New Mexicans, have proved the truth of that. The hardest task is to accept the truth of what experience has proved.

17

Wilderness

*In the long run we shall learn that there is no
such thing as forestry, no such thing as game
management. The only reality is an intelligent
respect for, and adjustment to, the inherent
tendency of land to produce life.*

—*Aldo Leopold*

HARD USE OF THE LAND changed society's economic relation
to the back country, and eventually it also changed most people's idea of what the purpose and meaning of wild lands
should be. From the late nineteenth century onward Americans increasingly viewed wild lands, not as a resource to be
tamed and exploited, but as a psychic refuge, a temple, and a
school for learning the traditional values of self-reliance and
love of nature. Predictably, society's change of heart was like
that of a fickle lover: wildness was not cherished until it was
almost lost, until it had almost vanished forever. Through the
boom years of stock raising and tie cutting, the empty spaces
of the West steadily shrank, and so did the fullness of life
within them. Fewer and fewer were the mountain sheep and
blue grouse; fewer were the tracks and scats and the haunting
primal sounds of the woods.

One of those sounds was the bugle of the elk. In the Pecos
country it was heard on frosty evenings in late September,
when cold tinged the aspen with gold and the rut of the bulls
began. Particularly the older bulls would advertise their am-
orousness by bathing in wallows of urine and mud, and at

sundown, reeking and randy, they serenaded their harems and challenged their competitors with a song that was as loud as it was vibrant. The bugle started out deep and low, something between a growl and a bellow, then rose up suddenly to a piercing whistle. It was a sound that carried for miles across the crisp parks of the high country, and its reverberations lingered in the canyons. The bull braced himself with stiff legs as he bugled; he lifted his head high until the ridge of his muzzle pointed to the sky and his antlers stretched over his back like flat-bottomed clouds above a plain.

But the fine sound of elk bugling was no longer heard in the Pecos country after about 1888. Market hunters, settlers, and miners had hunted the local herds to extinction. Another backcountry animal, the bighorn sheep, soon suffered the same fate. Although their natural range extended down to the edges of the plains, by about 1870 the sheep survived only among the highest peaks. There, because they obtained their nourishment, winter and summer, from relatively small patches of alpine grassland, they were especially vulnerable to competition from domestic sheep, which grazed those grasslands to exhaustion. Starvation, hunting pressure, and domestic sheep diseases like scabies gradually combined to eradicate the bighorn from the Pecos country, and tracks of the animal were last seen near the Truchas Peaks in 1903. Other species that followed down the trail to local or regional extinction included the white-tailed ptarmigan, whose principal food and cover, the alpine willow, was laid waste by herds of domestic sheep, and the pine marten, whose doom was ensured by the fine quality of its pelt. Timber wolves, too, which were never numerous in the high mountains, succumbed to hunting and trapping pressure early in the century and vanished from the area, eventually to be replaced by the wilier and more adaptable coyote. Greatest of all the mountain animals to be eliminated, however, was the grizzly bear, which, as deer and elk became scarce, turned to cow flesh and sheep. Guilty, like the wolf, of competing directly with man for food, the last of the Pecos grizzlies was hunted down in 1923.

Perhaps the most disheartening statistic from this period of

Photo 62. Bighorn sheep in the Pecos Wilderness, January 1978. Bighorns live all year above timberline, where the paucity of winter feed makes them extremely vulnerable to starvation and disease. Author photo.

wildlife devastation was that on the *one million acres* of the Carson National Forest only eight deer were killed during hunting season in 1915. Certainly hunters took home many more deer than that during the course of the year, for it was true then (and in some parts of New Mexico is still true) that virtually no man went into the woods for any reason without taking his rifle, and if he saw a deer, he shot it, regardless of its sex, age, or the time of the year. Nonetheless, a legal harvest of only eight animals from so vast an expanse of thinly inhabited country showed that the woods were empty and that the matrix of life within them had been largely destroyed.

Enlightened sportsmen around the state agreed that something should be done, and ultimately their concern led, in

1921, to the establishment of a state game commission that was relatively free of political control. At last, such hunting laws as there were would be impartially and effectively enforced. Hunting regulations on seasons and bag limits, however, could not alone restore the decimated populations of the state's game animals and offered no protection whatever to nongame animals. An intelligent plan for wildlife conservation also required the protection of wildlife habitat, and since so little of New Mexico's back country was privately owned, habitat protection ultimately necessitated a partial redefinition of the purpose of public lands.

It was no accident that the decade of the 1890s, when the federal government set aside the first Forest Reserves and National Parks (after Yellowstone), was also the first decade during which America lacked a wilderness frontier, at least in the forty-eight contiguous states and territories. Particularly in the heavily populated urban areas of the country, Americans became increasingly eager to preserve what they could of their wild heritage, and from the outset, preservation was a major goal of the National Parks. The Forest Reserves, soon renamed National Forests, on the other hand, did not emerge with a clear mission until Gifford Pinchot organized them around nonpreservationist, utilitarian purposes in the first decade of this century. Nearly another decade then passed before the forests were again the subject of preservationist talk, but when the debate revived, one of the loudest voices urging preservation came from within the Forest Service, and it belonged to a remarkable young New Mexico forester named Aldo Leopold.

Leopold had joined the Forest Service in 1914, and after a short stint on the Gila Forest in southwestern New Mexico, he moved north to serve on the Carson, first as Deputy Forest Supervisor, then as Supervisor from 1911 to 1913. A dedicated sportsman and hunter, Leopold lobbied energetically for more vigorous enforcement of the state's game laws and also for greater protection of the National Forests' rapidly deteriorating fish and wildlife resources. He pushed for increased Forest Service involvement in the enforcement of game laws, for the establishment of refuges, and for a measure he would later earnestly regret—the extermination of such major predators as wolves. After some early successes, Leopold found his superiors in the Forest Service too preoccupied with the imperatives of World War I to pursue the cause of wildlife conservation, and he resigned from the Forest Service and devoted himself to raising public support for his conservation ideas in Albuquerque. His intention was to rejoin the Forest Service as soon as the bureaucratic climate appeared more favorable.

That time came sooner than he expected. In 1916 the Forest Service lost a major Congressional battle when the National Park Service was created within the Department of the Interior to assume custodial responsibility for National Parks and

Photo 63. Deputy Forest Supervisor Aldo Leopold, left, at Carson National Forest Headquartes in Tres Piedras, 1911. Also pictured are Forest Supervisor C. C. Hall, right, and Forest Assistant Ira T. Yarnall. Photo by R. E. Marsh. Courtesy Forest History Society, Durham, N.C.

Monuments, many of which lay within National Forests. At issue was the future role that outdoor recreation would play in the life of the newly motorized American public. The Forest Service watched first with jealousy as the Park Service won instant popularity by promoting its lands as national playgrounds, and then with dismay and fear as the rival bureau made repeated overtures for the acquisition of National Forest land. Not to be outdone, the Forest Service resolved to enter the recreation business, and in 1919 Aldo Leopold rejoined the agency in order to break the necessary new ground in the Southwest.

Leopold's concept of managing the National Forests for recreation quickly went beyond the simple enhancement of hunting, fishing, and camping. He recommended that besides developing conventional recreation facilities, the Forest Service should also establish a system of what he called Wilderness Areas, each one being "a continuous stretch of country preserved in its natural state, open to lawful hunting and fishing, big enough to absorb a two weeks' pack trip, and kept devoid of roads, artificial trails, cottages, or other works of man." The main attribute of Leopold's Wilderness Areas was to be primitiveness, rather than unique landforms or beautiful scenery, as was the case with the National Parks. The distinction was a fine one, but very important. It meant that the Wilderness Areas would not be developed for the comfortable accommodation of automobile tourists, but would be left as rough and rugged as nature had made them.

For the rest of his life, both in the Forest Service and as a professor of game management and wildlife ecology at the University of Wisconsin, Leopold continued to refine his concept of the proper relationship between man and wildlands. But in Albuquerque in the early 1920s the immediate stimulus for his wilderness advocacy was the prospect that the backcountry of the Gila National Forest in southwestern New Mexico, which he knew intimately and valued as one of the last large roadless areas in the Southwest, faced imminent development. In 1921 he outlined a proposal for the creation of a 574,000-acre Wilderness Area within the Gila, and three years later after further study and review, Regional Forester Frank C. W. Pooler approved it. Soon other National Forest Wilderness Areas patterned after the Gila were established throughout the West, and the Forest Service at last found itself equal with the National Park Service as one of the nation's leading institutional champions of wild country preservation.

The Pecos High Country did not have long to wait before it, too, received similar administrative protection. The Forest Service formally recognized its exceptional primeval character and set it aside from most kinds of development by declaring it a Primitive Area in 1933. Twenty-two years later, somewhat to the amusement and amazement of the people who lived around its edges, the Pecos Primitive Area was redesignated the Pecos Wilderness.

To the people of the mountains a kind of retrogression of civilization seemed to be taking place. First a Reserve, then a Primitive Area, then a Wilderness—the high country was metamorphosing in ways no one had expected. No one had ever thought to call the high country a wilderness, mainly because in the then-accepted meaning of the word, it was not. It was backcountry, or high country, or simply *la sierra*, but it was not a wilderness any more than Santa Fe or Albuquerque was still a frontier town. The high country was conceived rather possessively as a large extended backyard—a commons—where people might make use of as many resources as the Forest Service would permit, and most people in the area had expected that sooner or later the high valleys would be laced with truck roads in order to open the country to logging. Some people even believed the local politicians who promised to have the government dig a tunnel through the East Divide and divert an entire river of irrigation water to Rociada, Sapello, and points east.

But instead of a tunnel, the local people received the new designation of Wilderness for their big backyard, and instead of logging roads, they had to wind their way through an ever-increasing proliferation of grazing and hunting regulations. Gradually, but not very happily, they realized that the mountain country stretching up and back from their ranches was

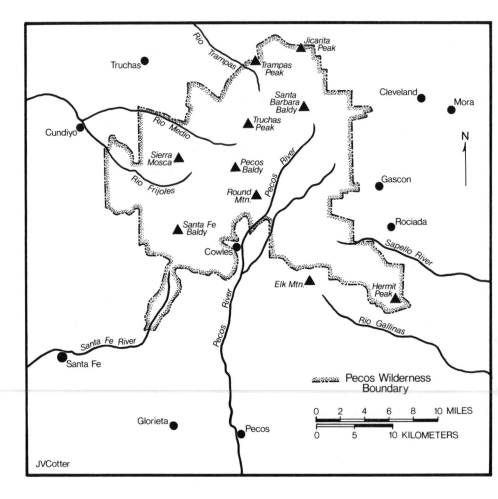

Map 8. Boundaries of the Pecos Wilderness, 1984.

no longer a "private" Hispanic commons, neither in the complete sense that their grandfathers had known, nor in the more conditional one of their fathers' time. Wilderness designation drove home the fact that the Pecos country was a grand, national kind of commons, the preservation of which was intended to benefit Texans, Easterners, and Californians as much as New Mexicans.

The *coup de grâce* to the image of the great backyard was not delivered until 1964. In that year Congress finally passed the Wilderness Bill, one of the most complex and far-reaching pieces of conservation legislation of all time. First submitted to Congress in 1957, the Wilderness Bill proposed the consolidation of all Wilderness Areas within a National Wilderness Preservation System and permanently prohibited any disturbance of their natural integrity. Aldo Leopold, who had died in 1948, was in many ways the spiritual progenitor of the measure. At every stage in the legislative process wilderness advocates repeatedly cited his ideas and writings, especially his call for the adoption of a "land ethic" to guide society in developing a harmonious and nondestructive relationship with the environment. David Brower of the Sierra Club went so far as to tell a Senate committee in 1961 that "no man who reads Leopold with an open mind will ever again, with clear conscience be able to get up and testify against the wilderness bill."

Brower's hyperbole notwithstanding, there was an enormous amount of negative testimony. A powerful coalition of logging, grazing, mining, and motorized recreation interests lobbied relentlessly against the bill and caused it to be rejected no fewer than sixty-five times. Altogether the legislative battle wore on for seven years, and the constant debate, the reams of testimony, and the public appeals made on behalf of the bill did much to popularize the virtues and benefits of wilderness recreation. By the time the bill finally passed, many thousands of Americans who had never before given a moment's thought to the idea of a Wilderness now wanted to visit one. They had come to see wild country as Leopold, and before him many others including John Muir and Henry David

[287]

Thoreau, had seen it: not as a storehouse of natural resources but as a refuge of unblemished peace and beauty where nature, working without interference from humankind, offered a partly ecological and partly spiritual enlightenment to those who came seeking it.

Appropriately, it was also in 1964 that the Pecos high country changed in a way that made it much more the wilderness it was touted to be. In that year Rocky Mountain bighorn sheep were finally re-established among the high crags of Pecos Baldy and the Truchas Peaks, a goal that until then had been obstructed by continued heavy use of the mountains by domestic sheep. Elk had by then been securely restored to the high country for several decades, and except for the enormous void left by the continued absence of the grizzly bear, the tracks and scats and sounds of the woods began to approximate what they had been a century earlier.

Photo 64. Elliot Barker packing elk trophy out of Upper Pecos, c. 1955. Courtesy Santa Fe National Forest.

Today we are obliged to think of our wildernesses as islands, whereas they used to flow together in a giant continental sea. They are the last remnants of a primal wildness that was both the fiend and muse of centuries of geographic conquest. But today's wildernesses are qualitatively as well as quantitatively different from the wilderness of the past. Their flora and fauna have undergone radical changes as a result of human activities. In spite of many people's nostalgic desires, the modern preserves are not and can never be "ecological wildernesses" in the sense of exhibiting the pristine vegetation and wildlife of pre-Columbian America. Indeed, defining the term *pristine* in any meaningful way becomes virtually impossible when one takes into account the landscape-changing habits of American aborigines.

The exotic plants—"green immigrants"—that have settled in the mountains include Kentucky bluegrass, sheep fescue, dandelions, various daisies, thistles, pasture grasses, and many other plants, all of which are now as much a part of the mountains' ecological community as any of the "native" plants that grow beside them. Similarly, the rainbow and German brown trout that occupy nearly every mountain stream cannot be persuaded to give their waters back to the native cutthroat they replaced. Whether the introduction of these exotics was intentional or accidental, beneficial or detrimental, their presence is an irreversible ecological fact.

In its contemporary context, the idea of wilderness is a sociological, not an ecological concept, and the Pecos and other modern wildernesses owe their preservation to legislation rather than environment. Essentially, wilderness is whatever people say it is: a park for dispersed, nonmechanized recreation, a biological research area, a gene preserve, a game preserve, an outdoor temple of solitude, or a museum of environmental history. Most wildernesses, including the Pecos, combine all those functions, but in order to do so effectively their caretakers must specify the objectives of their ownership and then *manage* the land in order to attain them.

There is no greater irony in the management of public lands today than the fact that the wildness of modern wilderness is to a large extent a manufactured and regulated quality. Game populations are manipulated by hunting seasons and bag limits. The activities of human visitors pose the greatest threat to the preservation of wilderness solitude, and often must be regulated or restricted. Even the ecological functioning of the wilderness requires human intervention in order to preserve the qualities the wilderness was created to protect. This last paradox is amply illustrated by the problem of managing fire in the wilderness.

As a natural component of wilderness ecosystems, fire periodically clears the way for the renewal of mountain forests and grasslands. A fire that rushes through a mountain meadow in late summer primes the grasses for vigorous growth the next spring and also destroys the conifer seedlings that continually invade the meadow from the sides, reducing its size. And a fire that burns away part of a forest makes possible the establishment of one of the mountains' colorful and highly productive stands of aspen. Although spruce and fir seedlings wither without partial shade, young aspen thrive in the hot, sunburnt conditions that follow a fire, and by reproducing by means of root sprouting as well as sexually through seed production, they can recover a charred landscape relatively quickly. All the large aspen stands of the high country from Santa Fe Ski Basin to the Santa Barbara Canyon occupy the sites of old burns, but without fire, the longer-lived spruce and fir that prosper in their shade will eventually grow up to replace them. The famously spectacular blaze of gold that the aspen wear in autumn is a visual echo of the flames that originally prepared their way and that are needed again to renew the stands and keep them vigorous.

Unquestionably, the harsh days of land abuse at the end of the last century featured far too many fires, most of them human-caused, but since that time the Forest Service, allied with Smokey, the most successful bear in the history of public relations, has defended the high country against *all* fire and has done its well-intentioned job so well that the mountain parks are smaller and the aspen fewer every year. Probably overgrazing also favored the expansion of forests into grass-

Photo 65. *Big burn on Santa Fe Baldy, thirty years after the fire. Portions of this area have been reclaimed by aspen, others by grasses. Photo by T. Harmon Parkhurst, 1916. Courtesy Santa Fe National Forest.*

lands by opening bare spots and removing dead plant material that kept conifer seeds out of contact with mineral soil. In the absence of fire, the Pecos country changed dramatically. One old forest ranger noted in 1953 that the high country "was a long ways from the open country it was in the 1920s." And since then the encroachment of the spruce-fire forests has not slackened.

In the late 1950s and early 1960s the Forest Service attempted to control the spread of conifers into productive rangelands. Crews armed with chainsaws set to work cutting young spruce and fir from the edges of high country meadows, but their labors, though expensive, released only a little grassland from the domination of the trees. Most of all, their efforts proved the inadequacy of mechanical substitutions for fire.

Today, sawn stumps may still be seen near many forest openings, particularly along the Rito Jarosa, but they are no longer at the dividing line between grass and forest duff. They rot in the shade of young conifers fourteen and fifteen feet high, while still younger trees continue the relentless march of forest into the meadow.

These changes create difficulties for wilderness management on a number of levels. The loss of open parks and aspen stands significantly reduces the scenic attractions of the high country, at least for most people. Although the interior of the forests can be impressive and inspiring, it is drearily monotonous compared to the dramatic vistas of the open spaces and the brilliant colors of the aspen. Similarly, although conifer forests afford valuable protective cover, they are a relatively sterile environment for wildlife since little but the scrawny whortleberry and a few mosses grows in their understory. Grasslands, by comparison, support a rich flora and fauna, and the "edge effect" created by the interfingering of meadow and forest is especially valuable to deer and elk. Aspen stands, too, owing to their relatively open canopies and the resultant lushness of their understory, benefit wildlife populations up and down the food chain from meadow mice to mountain lions.

The strongest argument of all for reintroducing fire to the wilderness, however, is the simple fact that it belongs there. If wilderness is "where the earth and its community of life are untrammeled by man," then natural processes should be allowed to take place to the fullest extent possible. But natural wildfires are not easily controlled. Because of long winters and abundant summer precipitation, conditions dry enough for a forest-renewing fire are rarely achieved in the high country, and when they are, they present a substantial danger to surrounding commercial timberlands, human habitations, and watershed stability. With only a slight change in wind speed and direction, a desired 300-acre burn can become a major conflagration with disastrous environmental and economic consequences.

One alternative to wildfire is the use of prescribed burning, which involves actually setting fire to grasslands or forest understory when the conditions of temperature, wind, and fuel moisture ensure that the fire can be properly controlled. The use of prescribed burning in a wilderness area would be richly ironic: forest managers would be deliberately setting fire to the land in order better to keep the land in a state "unaffected" by man. Prescribed burning also poses serious logistical problems for areas as inaccessible as the Pecos Wilderness, and since it violates most people's romantic notion of wilderness as land untouched by human influence, it may never be used. And so the meadows will continue to shrink and the aspen will slowly succeed to the dullness of spruce and fir.

The greatest dilemma associated with wilderness, however, is social, not ecological. Where once the backcountry was endangered by having too few friends, now it is threatened by too many. In 1911 Ranger Tom Stewart and his young assistant Elliot Barker, who later distinguished himself for twenty-two years as state game warden, estimated that no more than three hundred people visited the high country solely for the sake of recreation. By 1932 the number of hunters, fishermen, and pleasure campers had grown only to about eight hundred, and the mountains were still a relatively empty place where little happened but the weather and the yearly cycles of herding and roundup.

In 1948 two sturdy young men amused the old-timers of the high country when they sent their horses home from the top of the Santa Barbara Divide, made their gear into a pair of eighty-pound packs, and set out *on foot* for two weeks of hiking and fishing. Their choice of transportation was novel enough that *New Mexico Magazine* featured the trip in its July issue. In a land where "a man afoot was no man at all," backpacking provoked considerable astonishment, and that astonishment only grew as backpacking later became one of the most popular outdoor sports in America.

About 1960, backpackers came to outnumber horsemen in the high country, and their numbers continued to grow, both as a result of the new environmental consciousness arising from wilderness politics and as a result of technical improvements in the manufacture of durable, lightweight camping equipment. Many new aficionados of backcountry travel came to the Pecos Wilderness from Texas and Oklahoma, and some came from as far away as the East Coast. But the majority of the new wave of backcountry hikers came from the growing urban centers of New Mexico itself: Albuquerque, Santa Fe, Roswell, and Las Cruces. Their numbers grew with such speed that the Forest Service could not muster a reliable estimate of them until 1973 when it began to require permits for wilderness travel. The finding for that year was dizzying: 18,598 pleasure campers had tramped the trails of the Pecos Wilderness. In two years more the number passed 20,000, and the Pecos emerged as one of the five most heavily used large wildernesses in the National Forest System.

Photo 66. Backpacker on snowshoes approaching Pecos Baldy Lake, 1978. Today recreationists visit the high country in all seasons. Author photo.

In the high reaches of the wilderness one immediately began to see the toll of too many visitors at all the prettiest spots. Lakeshores erupted with an acne of campfire rings. The soils along the edges of popular fishing streams became compacted and barren, as one party after another camped in the same spot. Many campsites, worn down with heavy use, took on the look of a sandlot second base. Trails became rutted, firewood scarce, and the brilliant silence of the woods increasingly gave way to human noises. Recognizing that the press of humanity on the fragile ecosystems of the high country was causing serious environmental damage, the Forest Service prohibited camping in the most seriously overused areas. These closures, while never entirely effective, required a plethora of unfriendly signs and the regular presence of patrolmen. Clearly the gulf between *wilderness* and the reality of a modern Wilderness Area was growing wider—and it still is. The increasingly crowded trails and campsites of the high country all too often seem as shabby and unexceptional as the urban and suburban environments most wilderness visitors come to the high country to escape.

The stockmen who graze their cattle in the high country and their neighbors in the villages below watch the coming and going of the backpacker tourists with amusement and sometimes with resentment. To them the idea of hauling a heavy pack up a steep mountain trail suggests work much more than play, and it leads them to suspect that the reason the backpackers work so hard when they play is that they work so little the rest of the time. The backpackers, on the other hand, especially if they are unfamiliar with the cultural history of the region, tend to feel superior to the villagers. They notice that although the villagers live in the mountains, they rarely venture up to the picture-postcard lakes and peaks of the wilderness. Many tourists interpret this failure to mean that the villagers have little aesthetic appreciation of nature.

These are familiar stereotypes, and each contains a small kernel of truth. The villagers, like rural people everywhere, view their thinly settled environment from a more utilitarian perspective than city people. It is, after all, their home, and they and their ancestors have depended on it for their subsistence for centuries. This hardly means, however, that they do not appreciate the extraordinary scenery that surrounds them. The beauty of the mountain villages is a source of pride for all the people living in them, and a traveler in the southern Sangres who is fortunate enough to meet some of those people frequently hears the remark, "There is no place as beautiful as Valle," or Trampas, or Llano, Truchas, Mora, Rociada, or any of a score of other towns.

That same love of beauty is especially strong among National Forest grazing permittees, many of whom persist in ranching even though the increase of their tiny herds scarcely rewards their considerable labor. One reason they continue to take their animals to the high mountains is that they themselves wish to go there, and thanks to the animals, they can place their visit in a context of work, without which they would feel absurd and inauthentic—like tourists.

Not many subsistence farmers or ranchers have affinity for the role of tourist, least of all when the land they are expected to tour is land they consider their home. And logically, not many backpacking tourists ever succeed in understanding the land as feelingly as people whose traditions bind them to it. Nevertheless, in spite of the attitudes that divide them, each group has much to teach the other in the way it comprehends and uses the wilderness.

Wilderness preservationists, contrary to the accusations of their critics, are not attempting to reserve the backcountry for the exclusive use of a relatively few outdoor enthusiasts. Rather they contend that Wilderness Areas are, or should be, institutions for moral education. Wilderness experience, they say, educates people in the virtues of self-reliance, independence, self-restraint, and respect for nature, all of which qualities are not only useful but essential to survival in a world of shrinking resources and increasing complexity.

The mountain villagers of the southern Sangres have traditionally possessed the qualities of self-reliance and independence in abundance. Self-restraint and respect for nature, however, have been much less conspicuous. If there is a flaw in the relationship of the villagers to their environment, it is that they, like the people of pioneer and subsistence cultures everywhere, have consistently underestimated their capacity for injuring the land. The mesas and mountains may indeed be the *alma*, the soul of village culture, but their elevated status has not protected them from abuse. In the years to come, as the timber and mineral resources of northern New Mexico attract progressively greater attention from industry, the wilderness ethic of using the land nondestructively can provide villagers with an ideological focus both for their own use of the land and for influencing the Forest Service in its administration of the National Forests.

The backpacker tourists, meanwhile, may espouse that ethic intellectually, but many of them consistently fail to put it into practice, perhaps because they lack the complementary virtues of self-reliance and independence. Two by two and three by three, almost always in groups, they follow each other up the wilderness trails, always to the same splendid, scenic campsites, which their numbers gradually destroy. As their guidebooks admonish them, they "take only photographs and leave only footprints," but so many of their footprints landing in the same places leave the land seriously scarred. In their quest for the most picturesque alpine lakes and the highest, most dramatic peaks, they never discover what the cowboys and sheepherders of the past learned naturally: that one begins to understand the mountains only by understanding what is ordinary and usual about them, and one learns about those things in places away from the tourist throng.

So long as they are not overrun with people, the mountains of the Pecos Wilderness live up to the claims that writers in the tradition of Muir and Leopold have made countless times for wild country everywhere. They are claims that boast of serenity, peace, freedom, and many other good things, and they are as varied as the people who make them. If a list of the outstanding qualities of the Pecos Wilderness were to be made, two words that should certainly appear on it are continuity and endurance. It is remarkable, considering the abuse it has taken in the past, that the Pecos country is as wild and natural as it is. Backpacking tourists and mountain villagers can together rejoice that the mountain streams are no longer laden with silt from eroding hillsides and that there is more game in the woods now than at any time in the past century. Cattle, although in smaller numbers than formerly, still low in the meadows and parks, tended by stockmen whose fathers and grandfathers tended cattle in those valleys before them.

These continuities exist not because of inexorable natural processes but because people decided to restore and preserve them, and no such action has been more dramatic or more universally welcome than the restoration of the elk. On frosty September evenings the bugle of the elk echoes from ridge to ridge through the high country, as it did two hundred or two thousand years ago. It is a sound that harkens back to the depths of Pleistocene time, a call akin to the whoop of cranes and the howling of wolves. No other sound in the mountains is quite as brazen or as profoundly mysterious.

The first time I heard it, I followed the sound until I found thirty or forty elk scattered across a bowl-shaped meadow in the side of a ridge near Jarosa Creek, which, for lack of any other name, I call Jarosa Ridge. To the west, above the crest of the ridge, the sky shone a luminous rose, and so did the bottoms of the clouds that floated overhead. The cow elk and their calves fed in the shadows that spread across the meadow, while higher up, overlooking them stood the dominant bull. He had been wallowing and his sides were gray with mud. From time to time he erupted with a hungry coughing grunt and trotted downhill to nose a favorite cow, but then he returned to his higher vantage point and gathered himself to bugle. He bugled as though he meant to be heard from the Great Plains to the Rio Grande and the frosty air seemed brittle with his call. The younger bachelor bulls, who lurked around the edges of the herd, lifted their heads with every bellow, but they did not answer. They seemed content with their lower status—and with their prospects for *amour*, out of sight of the chief, at the edge of the trees.

Photo 67. The heart of the Pecos Wilderness, looking into the watershed of the Rito Azul. Photo by Harold Walter. Courtesy U.S. Forest Service.

CHAPTER 17

The concert continued as the red autumn sky burned down to embers over Jarosa Ridge, and gradually the meadow and the living forms upon it dissolved in darkness.

Later that evening a pack of coyotes crossed the ridge, singing on the run. They sounded shrill and possessed, as coyotes do, like lunitics gone rhapsodic with their madness. In a sense, that sound symbolizes a countercurrent to the ancient continuities of the wilderness, for a century ago there were no coyotes in the high country. It is a remarkable sound, and in its cacaphony and dedicated weirdness it suggests the sounds of other places far from the wilderness. I have wakened in the city to the sound of multiple sirens and mistaken them for the coyotes of home.

The elk, however, sing the song of the mountains. Their bugling is deliberate and full of spaces, yet nonetheless haunted and haunting. It seems to come out of a part of the world that, in the Pueblos' phrase, is raw—a part still forming, still unfinished. One of the greatest gifts of the wilderness is the knowledge that such rawness still exists. Whether in a mountain village or in the concrete canyon of a city, it is cheering to remember that the elk will bugle again in the mountains this fall and that every night of every year, the sun goes down over Jarosa Ridge.

18

Trails

*If these mountains die, where will our
imaginations wander? . . . And if the long-time
people of this wonderful country are carelessly
squandered by Progress, who will guide us to a
better world?*

—*John Nichols*

A TRAIL CUTS ACROSS Jarosa Ridge, wide and deep, worn down
to rocks and the roots of trees. This is the Gascon Trail, along
which for over a century the stockmen of Gascon, Rociada,
and Ledoux have driven their cattle to the summer ranges of
the upper Pecos watershed. The trail reveals much about the
history of man's relation to the mountain environment. In its
first miles it scales the steep wall of the East Divide along
treacherous narrow switchbacks. In spite of copious amounts
of dynamite and the best efforts of both permittees and the
Forest Service, there are places on the trail that cattlemen still
dread, stretches where one cow can hook or jostle another and
launch it on a fatal downhill tumble of hundreds of yards. You
understand that dread if you ride the trail, with its bad footing,
ledges, and the persistent horseflies that unsettle humans,
horses, and cattle alike. Near the top of the divide, however,
a fresh alpine breeze drives the horseflies away, and as you
clear the trees, the horizon leaps back a hundred miles in every
direction, except straight ahead where you see the naked gray
crags of the Truchas Peaks and the dark forest sea of the high
country. The thirsty cattle bellow as they trot forward, tasting

a hint of water in the cool thin air, and amid the excitement of the animals and the sudden explosion of space around you, you feel a rush of joy that is as much built into the trail as the danger and the steepness. The men who built the trail and drove the first herds over it felt that joy, and so do the cowboys who still drive cattle up the Gascon Trail.

Westward from the divide the trail broadens, frequently splitting into three or four deeply rutted parallel tracks that testify to the large numbers of animals that have been driven along it. On the west side of Jarosa Ridge and in many other places the heavy traffic has combined with thunderstorms and snowmelt to erode the trail into gullies, which in spite of efforts to stabilize them with logs and brush, have still not healed. Here is evidence of the former hard use of the land. Earlier in the climb you saw a reflection of the enterprise and resilience of the village ranchers. Now in the gullies and in the fans of soil they leave at the base of the slope, you see the negative side of the rugged mountain past. The trail helps to tie the two together.

All of the trails in the high country have a story to tell, and collectively they reveal much of the mountains' natural and human history. Each trail is like a sentence, a line of encoded intelligence stretching through the landscape and punctuated by the contours of the land itself. To read these trails is to learn something of the complexity and reciprocity that characterize the workings of the mountain environment. The first trails were the rivers and ridges laid out by water, wind, and the accidents of geologic history, and these are the trails that the elements still follow—and still carve. On them the mountain wildlife made another grid of pathways, and on top of that lay the grid, now largely erased, of the early people whom we know only by their dart points and grinding stones. And then more trails: the Pueblo network, the Apache, the Spanish.

Photo 68. The Southern Sangres from Skylab. Courtesy Earth Resources Data Center, Sioux Falls, S.D.

In every case the trails that humans blazed reflect both the needs that they brought to the mountains and the limitations that the mountains imposed on their efforts to satisfy those needs.

Some of the trails still in use in the mountains were blazed hundreds of years ago. The trail that connected Pecos Pueblo to the Tewa country before and during the colonial era descends into Santa Fe by way of Canyon Road, the center of the city's fashionable artists' colony. And the trail that hikers sometimes take south from the Rio Medio along the western foot of Sierra Mosca is the same one that Trucheros used to call their "secret tunnel," for it afforded them the safety and concealment of thick forests along a direct route to Santa Fe.

In the 1800s an expanding population and economy caused trails to proliferate in the mountains. There were new trails between villages, stock trails to new pastures, prospecting trails, and hunting trails. Anglos came to the mountains and built wagon trails to their remote homesteads, one of which connected Round Mountain, now in the Pecos Wilderness, to the ranches of the upper Pecos valley. Its ambitious builder believed that by raising racehorses in the high-altitude parks on top of the mountain, he could breed animals of superior stamina that would earn fortunes at the racetrack. Like many other mountain dreamers, he will be remembered less for his ideas than for his trails.

Later came modern trails. Regularized and systematic, a network of drained and culverted Forest Service trails leads campers from one end of the high country to the other, and lower down, in the work-a-day reaches of the National Forest, truck trails and logging roads connect the valleys and ridges with similar efficiency. The new trails replace the old ones in a process of succession that itself is a kind of trail through time. And twined into it are the trails of people's lives, the trails of George Beatty and Giovanni Agostini, of the Pecos survivors and the families of Picuris, of Vicente Romero and Juan Ortega, Frank Bond, Tom Stewart, and of tens of thousands of mountain villagers like Jacobo and Eloisa Romero.

All these trails, interwoven, make the fabric of the past, and however we value it, when we look over our shoulders, we are fortunate if we see it clearly.

The astonishing thing about the trails that thread through the history of the southern Sangres is their great length and continuity. None of the region's major culture groups has simply replaced or absorbed another, and in spite of the passage of centuries, none has lost its distinctiveness or authenticity. The Pueblo Indians still inhabit their ancient villages and celebrate the changing seasons with elaborate ritual. The Hispanos of the mountain villages still cling tenaciously to their language and traditions while they await a better economic day. And Anglo-Americans, latecomers to the region, have settled in as a powerful minority among their ancient mountain neighbors.

In great cities like New York and San Francisco one can find as much or more cultural diversity, and in isolated enclaves on some Indian reservations and in Appalachia or southern Louisiana one finds historical continuities as long or longer. But nowhere can one find diversity and continuity combined as richly as in northern New Mexico. The survival of the region's unique cultures is, in part, an accident of geography. Because New Mexico possessed scant mineral resources and few attractions of any other kind and because hundreds of miles of bleak and dangerous territory separated it from the economic centers of both New Spain and nineteenth-century Anglo-America, neither imperial nation undertook to colonize it with much energy or enterprise. Both Spain and the United States found that New Mexico was mostly unsettleable, at least with pre-industrial technologies. The arable lands of the territory consisted almost exclusively of narrow river and stream valleys and were too few and too widely scattered to allow the growth of large agricultural populations. Instead the landscape produced a dispersed pattern of settlement consisting of numerous small enclaves of population and culture. These Pueblo and Hispano villages became bastions of cultural preservation, for they were at once so self-sufficient that they had little need for the outside world and yet so poor that the outside world had little need of them. In isolation they persisted for centuries, changing little.

Because of the poverty and isolation of the region, the Spanish never developed a strong presence in New Mexico during the seventeenth century, and the Pueblos, when they at last achieved a semblance of political and military unity, were able to drive the Europeans out of the region. Although the Spanish returned to reconquer New Mexico a dozen years later, the Pueblo Revolt of 1680 taught them a high degree of tolerance for Pueblo customs and beliefs. Thanks also to the dangers posed by the hostile Apache, Comanche, Ute, and Navajo, the Spanish colonists learned to live as allies and neighbors with the Pueblos, rather than as conquerors exploiting them.

The qualities of isolation and poverty also deflected the empire-building energies of Anglo-Americans toward more richly endowed areas like California and Texas, which were also favored by access from the sea. Because of its impoverished and unsophisticated village culture, New Mexico proved as unimpressive to most Americans as it had to Spanish inspectors like Fray Domínguez. Albert Pike, an Arkansas journalist, summed up the feelings of many Mexicans and norteamericanos alike when he wrote in 1833 that "the Territory of New Mexico is the Siberia of the [Mexican] Republic." The persistence of such sentiments through the following decades helped to ensure that the Americanization of New Mexico proceeded extremely slowly.

The character of the land, however, does not alone account for the slowness with which Anglos settled the territory. As Lieutenant William Carpenter of the Wheeler Survey observed, "the emigrant wagon is . . . seldom seen bringing pioneers to new homes in this desirable region, while other Territories, far less inviting are being populated . . . rapidly." Westering Anglos were steering their wagons to Arizona, Colorado, and elsewhere, in part, because New Mexico was full of Mexicans, most of whom were Roman Catholic, possessed dark skins, and spoke Spanish, all of which from the point of view of nineteenth-century American Protestants gave evi-

in 1922 when Senator Holm D. Bursum of New Mexico introduced a bill in Congress that would have legitimized Anglo
and Hispano encroachments on Indian land and would have
relegated to hostile state courts the responsibility for adjudicating all future disputes concerning Pueblo land and water
rights. The actual author of this egregious legislation was Alois
B. Renehan, who had represented Frank Bond in the litigation
over the Las Trampas Land Grant. Collaborating with Secretary of Interior Albert B. Fall, a New Mexican who would soon
disgrace himself and the Harding administration in the Teapot
Dome scandal, Renehan devised a law that would deprive the
Pueblos of as much as 60,000 acres of their most valuable
irrigated land and would also place them at a permanent dis-
advantage in defending their legal rights. The Pueblos clearly
recognized the desperateness of their situation. In an appeal
for help to "the People of the United States," the delegates of
the newly formed All Pueblo Council charged that the Bursum
Bill would destroy their people's "common life and will rob
us of everything we hold dear—our lands, our customs, our
traditions."

Many Anglos rallied to the Indians' defense, foremost among
them the artists and intellectuals who had settled in Santa Fe
and Taos since the turn of the century. Drawn to the region
by the beauty of its landscape and the stimulation of its diverse
cultures, they were idealistic, eccentric, and often wealthy.
The most prominent figure among the Taos intelligentsia was
Mabel Dodge Sterne, a New Yorker who moved west in 1917
and eventually married Antonio Luhan, a Taos Indian. It was
Mabel Dodge who brought D. H. Lawrence to Taos and it was
also she who enlisted John Collier, a young New York City
poet and social worker, to work on behalf of the Pueblos.

Collier proved to be an effective advocate of Indian rights.
Not only did he contribute materially to the defeat of the
Bursum Bill, he also lobbied successfully for the restoration
to the Pueblos of lands that had been encroached upon by non-
Indians. Eventually some 667,479 acres were returned to Pueblo
ownership, with their Anglo or Hispano "owners" being com-
pensated by the federal government. In 1933 Franklin D.

Roosevelt appointed Collier Commissioner of Indian Affairs, a position he held for twelve years. Collier's leadership and the growing political sophistication of Indians throughout the country helped bring about the creation of the Indian Claims Commission in 1946. For the next thirty-two years the Commission reviewed and resolved a large number of serious land disputes concerning the Indians of the United States.

The lands that have been restored to the Pueblos have significantly increased the tribes' economic resources. They have also widened the geographic barriers that protect the Pueblos from the pressures of the outside world, and they have added immeasurably to the sense of cultural pride and independence of contemporary Pueblos.

The Hispanos of New Mexico, however, have not been so fortunate. In many cases they were deprived of their lands through outright dishonesty or through legal tactics, like the suit for partition, that violated the spirit of the Treaty of Guadalupe Hidalgo. In spite of these historic abuses, however, the Hispanos have never received the level of protection that was accorded their Indian neighbors. Students of New Mexican affairs as diverse as Reies Tijerina and Donald Cutter, who for many years chaired the history department of the University of New Mexico, have long urged that it is still not too late to redress old wrongs and that a good first step toward that end would be to create a Hispanic Land Claims Commission modeled on the Indian Claims Commission.

Photo 69. Jury for Trial of a Sheepherder for Murder by Ernest L. Blumenschein, 1936. A masterwork in strictly visual terms, this painting by one of the founders of the Taos art colony suggests the uneasiness felt by many villagers in dealing with Anglo law and society. Note the faceless "Father of Our Country" at top center. Courtesy Museum of Western Art, Denver.

However successful such a commission might be in read-judicating land grant claims, it would be a poor gamble to stake one's hope for a better future on such an idea. Widespread restitution of land grants to the descendants of the original grantees is, above all, a political impossibility, for it would cast doubt on the legitimacy of the ownership of vast tracts of land from the Gulf of Mexico to the Pacific Ocean. And even if full restitution were possible, it would be no panacea for the problems that assail Hispanic villagers today. The timber and range resources of the grants, which the Forest Service has improved at considerable cost to the taxpayers of the United States, are simply not productive enough to support even a fraction of their putative inheritors. At best they might yield a modest annual dividend to their owners and provide jobs for a number of villagers slightly greater than the number who work for the Forest Service today. The difficulty of managing the grants, however, would be daunting, to say the least To apportion ownership rights equitably between present residents and the many legitimate heirs who have long since moved away would tax the wisdom of a thousand Solomons. And beyond that, to create and maintain a workable consensus on management goals would require an extraordinarily high level of political skill and leadership.

Under any system of ownership, there would need to be safeguards in the management of the grants to prevent the kinds of abuses that have been so typical of the past and so destructive, in their own way, of village culture. A return to stock raising on the scale practiced during the first third of this century would only produce new cycles of erosion and further losses in the biological productivity of mountain lands. The restoration of the grants to the descendants of the original owners, in fact, might ultimately be somewhat ironic, for if the principle of sustained yield continued to guide land use, then the new owners of the grants might surprise themselves by how little they departed from the major management practices of the Forest Service.

Fortunately, the readjudication of land grant claims is not

the only or necessarily the best means for redressing the griev-ances of Hispanic New Mexico. One alternative, while more moderate, is considerably more realistic, for the framework for pursuing it is already in place. Admittedly, this alternative sidesteps the issue of who really owns the land, but it pos-sesses considerable potential for practical benefit. It involves taking part in what at first glance might seem a tiresome and bureaucratic endeavor: the land management planning process for the National Forests.

This might be an ineffectual way to invest time and energy were it not that the Forest Service takes planning very seri-ously, thanks in large measure to the National Forest Man-agement Act of 1976. This law requires that each National Forest develop a comprehensive land management plan every five years, and that the planning process "provide for public participation in the development, review, and revision" of its ultimate product. To be sure, this elaborate process has proved to be time-consuming and expensive, far more so, in fact, than even the most pessimistic original predictions. Many initial planning problems, however, should not recur since they are associated with the accumulation of the extremely large amounts of baseline data that are the foundation of each plan. In principle, the planning process is sound, and the Forest Service remains strongly committed to it.

From the public's point of view, the most important stage in the articulation of a plan is the so-called scoping process, wherein the issues that a plan must address are identified and defined. Once an issue has formally been identified, the Forest Service is obliged by law not to disregard it in the formulation of its plan. This means that by involving themselves in the planning process, villagers and others can guarantee that their concerns will be explicitly addressed.

Extensive planning of this general kind is not new to the Forest Service, and neither is the involvement of the public in scoping and other functions. What distinguishes that NFMA planning process from its predecessors, however, is that where public participation was formerly a privilege, now it is a right.

This is not to say that the public's suggestions must be accepted or obeyed, but they must be taken into account.

The NFMA planning process has the potential to produce striking results, as exemplified by the draft plan for the Santa Fe National Forest, which was released in 1981. Like all NFMA plans, this one calculated the effects of forest management under a variety of alternative management goals and then recommended a preferred alternative. Usually the alternatives developed for these exercises simply reflect a range of commodity outputs, from, say, high volumes of timber and few recreational visitor-days to abundant recreation and little timber. In this case, however, the forest planners, on their own initiative, explored a management alternative pursuing a more original goal: the return of maximum benefits from forest activities to local communities. Even more significantly, this option was selected as the preferred alternative as summarized in the following statement of mission:

Manage the Resources of the Santa Fe National Forest under multiple use and sustained yield principles to provide for maximum contributions to the national welfare and to the economic and social needs of the people of Northern New Mexico. Management programs are to be oriented to maintain and enrich cultural values and a viable rural economy.

Maintain an on-going awareness by Forest Service personnel of the uniqueness of management associated with the National Forest in Northern New Mexico. Forest officers must recognize that local people rely on that land for a social and economic base and will continue to do so, and that their way of life is directly affected by the management of National Forest lands.

Because of errors in the computation of its data base, the 1981 draft plan was never accepted as final. Its successor, however, may well incorporate major aspects of the 1981 plan, including its preferred alternative. In fact, the people of northern New Mexico should see that it does, and that it devises ways to maximize benefits to local communities even more diligently and aggressively than the draft plan did. This is what informed and responsible participation in the planning process can lead to. The opportunity to participate should be seen as nothing less than a form of political enfranchisement.

Some pages back I said that in northern New Mexico the Forest Service is a more powerful political entity than county government. I think this is an important truth in the life of the region, and it raises an important problem for a political democracy like ours. Because the Forest Service bears major responsibility in a highly technical area, namely the protection of the ecological health and productivity of the land, its management decisions should be scientifically sound. But because it exerts so much influence over so many lives, its decisions should also be democratically acceptable. Unfortunately, the chain of accountability that leads through the state's congressional delegation of the Department of Agriculture and thence down through the ranks of the Forest Service to the local level is too circuitous for the efficient resolution of local problems.

The NFMA planning process offers an alternative channel for political communication without sacrificing scientific accountability. Properly used, it will make a new contribution to the functioning of democracy, albeit in an unfamiliar guise. Using it well, however, is hard work. There are no shortcuts to understanding the process, knowing the facts, and getting involved.

Among the many reasons why the people of northern New Mexico must work diligently to gain greater control over their own lives is one that is often overlooked: the Anglo invasion of 1846 is still going on; the Land of Enchantment is changing rapidly. As the trails forged by the Pueblos and Hispanos stretch onward, newcomers to the Sangres are continually blazing new trails beside them. For a brief time during the late nineteen sixties and early seventies, Taos and the surrounding mountains were a mecca of the countercultural world. Hippies from every corner of the country crowded to the Sangres seeking open space, freedom from a society they considered corrupt and repressive, and an opportunity to experiment with new social and personal identities. By 1969 there were close to twenty-seven communes active in Taos County, much to the dismay of most local residents. Fights and feuds between hippies and Hispanos became almost commonplace for a time, but fortunately the tide of the hippie migration soon ebbed. By 1973 only about four communes were still active, and many of the young Anglos who stayed on in the area settled down to a less ostentatious and iconoclastic way of life.

The flowering of the counterculture had a lasting effect in the southern Sangres, for it brought Anglos permanently into communities where none had lived before. Today at least one or two Anglo families live in virtually every mountain village, and many of them possess the same qualities of hardiness, self-reliance, and inventiveness that characterized the homesteaders of a hundred years ago. In some villages they have won the respect and friendship of their Hispanic neighbors. In others, relations are more often strained and tense.

It is yet another irony of modern times that the geographic isolation and perpetual poverty of the mountain villages, which for so long shielded them from the most direct effects of social change, now actually invite the penetration of the mountain world by outsiders. A harbinger of this turnabout was the location of the nation's first nuclear weapons research center at a former boys' school in the Jemez Mountains. Los Alamos, where the world's first atom bombs were built, was chosen as the headquarters for the top-secret Manhattan Project specifically because its isolation and inaccessibility ensured tight security. Since those days the town of Los Alamos, still dominated by the Los Alamos Scientific Laboratory, has matured into an extremely affluent Anglo community. It is an outpost of mainstream uniformity on an often tense cultural frontier.

In the years since World War II legions of Anglo-Americans have come to northern New Mexico from other regions of the country, although their reasons tend to be more like Mabel Dodge's than Robert Oppenheimer's. They come for the sake of a salubrious climate, spectacular surroundings, and a cultural milieu so diverse and exotic that all the rest of the country seems monotone by comparison. They have come in such

Photo 70. High country trail through aspen. Author photo.

numbers that the seams of Taos and Santa Fe are bursting with new immigrants. These new New Mexicans include the restless young and the restless old, and many of them openly claim to be temperamentally ill-suited for service in the family business "back east." Recognizing this, one pundit has argued that much of what is peculiar about Santa Fe can be explained by the fact that its affairs are so frequently controlled by "The Sons and Daughters Who Didn't Fit In."

Certainly the Sons and Daughters have taken over the heart of Santa Fe where the old comfortable plaza, once lined with drugstores, lunch counters, and other ordinary places, now supports a swarm of semikitsch art galleries and trendy boutiques. The descendants of Vargas's original colonists must now take their more mundane business to the proliferating shopping centers, ringing the city, that are just like shopping centers everywhere. In Taos, too, the Sons and Daughters are steadily taking over, thanks mainly to the economic power of their higher-than-average incomes, and they threaten to turn that town, in novelist John Nichols's words, into "just another paranoid, money-grubbing boardwalk—a mountain Las Vegas, Atlantic City, Tahoe, in miniature."

While adding to the cultural diversity of the region, the influx of Anglos to the cities, towns, and villages of northern New Mexico also threatens to weaken further the traditional Hispanic and Indian cultures. The rising cost of taxes and land and the homogenizing influence of consumer economics all conspire to speed the demise of old ways and old values. Along with the need to achieve justice in the distribution of land and resources, the greatest challenge facing northern New Mexico is to find a way to accommodate its immigrant populations while at the same time protecting and preserving the cultures it has inherited from the past. Perhaps the two challenges are really the same. Certainly the people of the region cannot meet one of them without also addressing the other.

The broadest considerations of New Mexico's future compel acceptance of these challenges. The traditional cultures of the region are not mere aesthetic ornaments to lure tourists and stimulate the imagination. Like a rare plant or a unique eco-system they are irreplaceable. Society needs to conserve them, both for the answers they give to the basic problems of existence and for the fresh new questions they pose about the proper relation of people to each other and to the land.

Appendix

Key to Map 5: Land Grants in North-Central New Mexico over 600 Acres. Adapted from unpublished tables provided by State Records Center and Archives and from *New Mexico in Maps*, ed. Jerry L. Williams and Paul E. McAllister. N.B.: Only grants confirmed by the United States are shown.

Map No.	Name of Grant	Year Issued	Acreage Confirmed
Spanish Indian Grants			
1	Picuris	1689	17,461
2	San Juan	1689	17,545
3	Santa Clara	1689	17,369
4	San Ildefonso	(document lost)	17,293
5	Pojoaque	(document lost)	13,520
6	Nambe		13,590
7	Tesuque	(document lost)	17,471
8	Cochiti	1689	24,257
9	Santo Domingo		92,398
10	San Felipe	1689	33,645
11	Pecos	1689	18,763
12	Santa Ana		17,361
13	Zia	1689	17,515
14	Jemez	1689	17,510
15	Sandia	1689	22,883
16	Taos	(document lost)	17,361
Spanish Grants, 1650–1800			
17	Piedra Lumbre	1766	49,748
18	Juan José Lobato	1740	205,616
19	Plaza Colorada	1789	7,578
20	Plaza Blanca	1789	8,955

Key to Map 5 (continued)

Map No.	Name of Grant	Year Issued	Acreage Confirmed
21	Abiquiu	1754	16,708
22	Polvadera	1766	35,761
23	Antonio de Abeyta	1736	721
24	Ojo Caliente	1793	2,245
25	Antonio Martínez	1716	61,605
26	Antonio Leroux	1742	56,428
27	Gijosa—Rancho de Taos	1715	16,365
28	Cristóbal de la Serna	1710	22,233
29	San Fernando de Taos	1799	1,817
30	Rancho del Río Grande	1795	91,813
31	Santa Barbara	1796	31,085
32	Las Trampas	1751	28,132
33	Sebastian Martin	1712	51,388
34	Black Mesa	1743	19,171
35	Bartolomé Sánchez	1707	4,470
36	Santa Cruz	1694–1706	4,567
37	Santa Domingo de Cundiyo	1743	2,137
38	Nuestra Señora del Rosario San Fernando y Santiago	1754	14,787
39	Francisco Montes Vigil	1754	8,254
40	Juan de Gabaldón	1752	10,690
41	Santa Fe	c. 1660	17,361
42	Jacona	1702	6,953
43	Caja del Río	1742	66,849
44	Ramón Vigil	1742	31,210
45	Cañada de Cochití	1728	19,113
46	Ojo de San José	1768	4,340
47	Cañon de San Diego	1798	116,287
48	San Ysidro	1786	11,477
49	Ojo del Borrego	1768	16,602
50	La Majada	1695	54,404
51	Mesita de Juana López	1782	42,023
52	Cieneguilla	1693	3,203
53	Sitio de Juana López	c. 1750	1,109

Key to Map 5 (continued)

Map No.	Name of Grant	Year Issued	Acreage Confirmed
54	Los Cerrillos	1788	1,479
55	Sebastian de Vargas	1728	13,434
56	Cañada de los Alamos	1785	12,068
57	Bernalillo	1742	3,404
58	San Antonio de las Huertas	1767	4,764
59	Alameda	1710	79,346
60	Bernabé Mantano	1753	44,070
61	Elena Gallegos	1716	35,085
62	Atrisco	1768	82,729
63	Pajarito	1746	28,725

Spanish Grants, 1800–1820

Map No.	Name of Grant	Year Issued	Acreage Confirmed
64	Antonio Ortiz	1819	163,291
65	Los Trigos	1814	7,342
66	Alexander Valle	1815	1,202
67	SanCristóval	1815	81,031
68	Santa Rosa de Cubero	c. 1805	1,945
69	Ojo del Espiritu Santo	1815	113,141
70	San Joaquín—Cañon de Chama	1806	1,423
71	Arroyo Hondo	1815	20,629
72	Lamy	1820	16,547

Mexican Grants, 1821–1846

Map No.	Name of Grant	Year Issued	Acreage Confirmed
73	Tecolote	1824	41,123
74	Tierra Amarilla	1832	524,215
75	Sangre de Cristo	1843	1,038,195
76	Beaubien & Miranda (Maxwell)	1841	1,714,764
77	Mora	1835	827,621
78	John Scolly	1843	21,701
79	Las Vegas	1835	431,654
80	Petaca	1836	1,392
81	Baca Location No. 1	1835	90,426
82	Ortiz Mine	1833	69,458
83	Tejón	1840	12,801
84	San Pedro	1839	20,094

Notes

INTRODUCTION: PLACE

Page

3   "east of the Villa . . ." Kessell, *Kiva, Cross, and Crown*, pp. 230–31.

4   identified on maps   Wheat, *Mapping the Transmississippi West*, 1:119.

4   cemented in the public mind   Pearce, *New Mexico Place Names*, pp. 144–45.

1. SPIRIT HOMES

15   "To say 'beyond the mountains' . . ."   Momaday, *House Made of Dawn*, p. 38.

17   control the environment   Richard I. Ford, "An Ecological Perspective on the Eastern Pueblos," in Ortiz, *New Perspectives*, pp. 3–6.

17   Cosmos, being orderly   Alfonso Ortiz, "Ritual Drama and Pueblo World View," *New Perspectives*, p. 143.

18   Tewa . . . communicate with speakers   Kenneth Hale and

David Harris, "Historical Linguistics and Archeology," in Ortiz, *Handbook*, p. 171.

18   more faithful than Cushing   Marc Simmons, "Pueblo History Since 1821," *Handbook*, pp. 218–19.

20   confess ignorance   Ortiz, "Ritual Drama," *New Perspectives*, p. 136; Eggan, "Summary," ibid., 298–99; J. Bodine, "Acculturation Process and Population Dynamics," ibid.

20   shrines, Tsin, Tsave Yoh   Ortiz, *Tewa World*, pp. 19–20, 77, 21.

21   sacred mountains   Ibid., p. 16. Parsons's version of this myth is essentially the same except that six instead of four directions are invoked, the two additional directions being up and down. *Pueblo Indian Religion*, p. 251.

21   Cochiti, and others   Ortiz, *Tewa World*, p. 19; Douglass, "A World Quarter Shrine of the Tewa Indians," pp. 159–73. In order to be consistent with the English names of the other sacred mountains, I have spelled Chicoma following general usage as reflected on most regional maps. Ortiz uses the Tewa spelling, *Tsikomo*.

21   San Antonio Peak   Harrington, *The Ethnogeography of the Tewa Indians*; Curtis, *The North American Indian*, 17:44–45; Hewett and Dutton, *The Pueblo Indian World*, p. 34.

22   Lake Peak   Ortiz, *Tewa World*, p. 141.

22   visited the Jicarita   Harrington, *Ethnogeography*, p. 339. Harrington depended exclusively on maps prepared by the Wheeler Survey (see Ch. 11) for the actual location of the landforms his informants named. Since the Wheeler maps of the southern Sangres are inaccurate, many of Harrington's data are contradictory and difficult to interpret. The legacy of Wheeler's imaginary geography does not end with Harrington. Vincent Scully drew much of his material directly from Harrington, and his work is similarly encumbered with geographic error.

22–23   less *seh t'a*   Ortiz, *Tewa World*, p. 129.

23   represented in effigy   Parsons, pp. 172, 214.

23   continually circulating   Ortiz, *Tewa World*, p. 19.

24   resonate visually   Scully, *Pueblo: Mountain, Village, Dance.*

CHAPTER 2

26 any number of small shrines Starr, "Shrines Near Co-
    chiti," pp. 219–31; Parsons, pp. 194–95.

26 Great Sand Dunes Hewett and Dutton, pp. 23–24.

26 entered the lakes Ortiz, *Tewa World*, pp. 19, 97.

27 "so that no Mexican . . ." Curtis, 17:70–71.

27 "mating ceremonies" Hewett, *Ancient Life*, p. 96.

27 cultural apocalypse Collins, "Battle for Blue Lake"; Marc
    Simmons, "History of the Pueblos Since 1821," *Hand-
    book*, p. 216.

29 Navajo . . . Comanche Parsons, p. 252.

29 sacred mountains Curtis, 1:90–91.

29 territory overlapped D. A. Gunnerson, *The Jicarilla
    Apaches*, p. 243.

30 "married couple" Opler, "Jicarilla Apache Territory," pp.
    313, 315. See also Opler, *Myths and Tales*, p. 243.

30 close to Taos Pueblo Opler, "Jicarilla Territory," p. 315
    and *Myths*, p. 47.

30 Gallinas Canyon Ungnade, "Archaeological Finds," p. 20.

2. INTO THE MOUNTAINS

31 "spirit of energy" Parkman, *The Oregon Trail*, p. 167.

31 protected themselves against drought Richard B. Wood-
    bury and Ezra B. W. Zubrow, "Agricultural Begin-
    nings," *Handbook*; Wendorf, "Archaeology of
    Northeastern New Mexico," pp. 57–59.

34 use of fire not . . . haphazard Lewis, "Indian Fires." Also,
    O. C. Stewart, "Forest and Grass Burning in the Moun-
    tain West."

35–36 krummholz Parr, "Development of Tree Islands."

37 located in cold windy saddles Wendorf and Miller, "Arti-
    facts from High Mountain Sites," pp. 40–43. See also
    Ungnade, "Archeological Finds in the Sangre de Cristo
    Mountains."

37 about the time of Christ Ibid., p. 49.

38 Rio Grande Valley . . . sparsely settled Dickson, *Prehis-
    toric Pueblo Settlement Patterns*, pp. 14–15. See also
    Linda Cordell, "Eastern Anasazi," in *Handbook*.

38 progenitors of Anasazi Cynthia Irwin-Williams, "Post-
    Pleistocene Archaeology" in *Handbook*, p. 35.

38   agriculture much more reliable   Dickson, p. 72; Woodbury and Zubrow, p. 55; Cordell, pp. 134–35.

39   indigenous to the valley   Richard I. Ford, Albert H. Schroeder, and Stewart L. Peckham, "Three Perspectives on Puebloan Prehistory," in Ortiz, *New Perspectives*, pp. 25–36.

39   greater emphasis on hunting   Ibid., p. 34. The problem with this argument is that if the Puebloans hunted bison with regularity and success, the herds would soon move beyond the Puebloans' effective hunting radius, creating a "buffer zone" between the hunters and the game.

39   new centers of population   Cordell, p. 145; Kessell, *Kiva, Cross, and Crown*, p. 10; D. Brown, "Structural Change at Picuris Pueblo," p. 20.

40   horseless bison hunter   Kenner, *A History of New Mexico–Plains Indian Relations*, pp. 5–8.

40   Pueblos in the high country   Miller and Wendorf discovered the potsherds, and their find was later supplemented by Ungnade, as described in their respective articles, cited above. Pete Tatschl of the U.S. Forest Service, Pecos Ranger District, discovered the intact pot on the Padre (Pete Tatschl, personal communication, 9/75).

3. Brothers on the Faultline

43   "It is pitiful to view . . ."   Villagrá, *History of New Mexico*, p. 42.

43   Pecos grew rich   Kessell, *Kiva, Cross, and Crown*, p. 12.

44   European diseases   Simmons, *New Mexico*, p. 65. See also, Crosby, *The Columbian Exchange*, Ch. 2, "Conquistador y Pestilencia."

44   Pecos and Picuris were unsurpassed   D. Brown, "Structural Change at Picuris Pueblo," pp. 21–22; Kessell, pp. 8–10, 489.

46   a tourist attraction   Ibid., p. 476.

46   Wichita Indians in central Kansas   Bolton, *Coronado, Knight of Pueblos and Plains*.

47   "most if not all the men . . ."   "*Memoria* of the Castaño de Sosa Expedition," as quoted by Kessell, p. 55.

Chapter 3

47   Development of the mines   Bannon, *The Spanish Bor-
      derlands Frontier, 1513–1821*, pp. 28–29.

48   fled down the Rio Grande   Twitchell, *The Leading Facts
      of New Mexican History*, I:330. Oñate condemned a
      number of the deserters to death in absentia. Although
      most of those who fled were forcibly returned to San
      Gabriel, it is not clear if the sentences were carried
      out. See also, Bannon, pp. 6, 79.

48   baptized more than seven thousand   Kessell, p. 92.

48   proprietary charter revoked   Ibid., pp. 93–95.

49   Peralta ultimately imprisoned   Ibid., p. 97.

50   "a sixteenth century . . ."   Ibid., p. 127. See also, Hayes,
      The Four Churches of Pecos, pp. 19–35.

50   valuable haul of hides   Kessell, pp. 99, 188–89.

50   Franciscans raided Kivas   Bowden, "Spanish Missions,
      Cultural Conflict, and the Pueblo Revolt of 1680," pp.
      220–23.

51   "the most indomitable and treacherous . . ."   Benavides,
      The Memorial of Fray Alonso de Benavides, p. 25. See
      also, D. Brown, p. 29; Kessell, pp. 150, 162.

51   "perished of hunger . . ."   Fray Juan Bernal, as quoted by
      Kessell, p. 212. See also Kenner, p. 18; Twitchell, *Lead-
      ing Facts*, I:343–49.

53   "their desertion of our holy religion . . ."   as quoted, A. B.
      Thomas, *After Coronado*, p. 59.

54   ransom the Picuris   D. Gunnerson, *The Jicarilla Apaches*,
      pp. 122–23. Thomas, *After Coronado*, p. 59. The exact
      location of El Cuartelejo has been the subject of con-
      siderable debate. See discussion by Kessell, p. 544, n.
      2.

54   Picuris who returned to Santa Fe   D. Brown, p. 35.

54   less than four hundred   Ibid., pp. 31, 34–35.

54   worst epidemic   Kessell, p. 163.

54   forced them to resettle Galisteo   Schroeder, "Shifting for
      Survival," p. 251.

55   reducing the carrying capacity   D. Brown, p. 23.

55   increasingly sedentary   J. Gunnerson, "Apache Archae-
      ology in Northeastern New Mexico," pp. 23–28.

56   population of the pueblo   D. Brown, "Structural Changes,"
      p. 47 and Brown, "Picuris Pueblo," in Oriz, *Handbook*,
      pp. 271, 275.

56   Pecos factions struggled   Kessell, pp. 227, 230, 295.

56   every military campaign   Ibid., pp. 360–64, 372–73.

57   appearing first in Taos   Kenner, p. 28.

57   sought the protection of Spanish arms   J. Gunnerson, "Apache Archaeology," p. 38; Kenner, p. 33. By the time of the American occupation of New Mexico, the Olleros had established themselves west of the Rio Grande near Abiquiu, and the Llanero band spent much of its time in various camps along the east slope of the Sangres from Taos to Truchas (D. Gunnerson, *Jicarilla Apaches*, pp. 160–65).

57   appeared as plaintiffs   Ibid., pp. 240–43.

57   wrath of the Comanches   Kessell, pp. 372, 385.

58   "do not last even . . ."   Domínguez, *The Missions of New Meixco, 1776*, p. 213.

58   outbreak of smallpox   Kessell, pp. 348, 378, 543, n. 60.

58   the Pecos were good carpenters   Domínguez, p. 214; Chavez, "The Carpenter Pueblo"; Kessell, pp. 132–33.

59   forty Indians remained   Ibid., 492.

60   trekked to Jemez   Hewett, "Studies on the Extinct Pueblo of Pecos." The refugees and their descendants, however, did not break all their connections with the Pecos area. For a time at least they perpetuated ceremonies associated with a deep limestone cave at Terrero, New Mexico, on the Pecos River between Pecos and the Wilderness boundary. The cave, which provides the setting for one of the most dramatic scenes in Willa Cather's *Death Comes for the Archbishop*, had long been sacred to the Pecos. In 1904, Edgar Hewett learned from the Indians that they had made their visit in August 1903 and had found everything in the cave to be all right (ibid.). In 1929 thirteen horsemen rode from Jemez to Terrero and stayed three days in the cave. The most recent recorded use of the cave appears to have been in 1943 when nine Jemez braves made the trip (New Mexico Geological Society, *Guidebook of Southwestern Sangre De Cristo Mountains*, p. 141).

4. HOLDING ON

61   "Taking by the hand . . ."   Court of Private Land Claims,

CHAPTER 4

March Term, 1893. Claim No. 225, "Act of Possession."

62   train of creaking carts   Moorhead, *New Mexico's Royal Road*, p. 49.

62   between 7,000 and 8,000 Hispanos   Zeleny, *Relations between the Spanish-Americans and Anglo-Americans in New Mexico*, p. 25.

63   French to attempt expansion   A. B. Thomas, *Forgotten Frontiers*, pp. 386–87; Bannon, *The Spanish Borderlands Frontier*, pp. 126–30, 169.

63   wait for *five months*   Domínguez, *Missions of New Mexico*, pp. 270, 277.

64   "a state of miserable panic"   Ibid., p. 270.

65   flow of arms   Luis Navarro García, "The North of New Spain as a Political Problem in the Eighteenth Century," in Weber, *New Spain's Far Northern Frontier*, p. 207. Also, Domínguez, p. 252.

65   unhappy settlers of Abiquiu   Swadesh, *Los Primeros Pobladores*, pp. 35–36.

65   Anza . . . passed through . . . in 1779   A. B. Thomas, *Forgotten Frontiers*, p. 124.

65–66   Embudo, . . . El Nacimiento . . . and other settlements   Ibid., pp. 88–89, 100–101.

67   Las Trampas and Las Truchas   Twitchell, *Spanish Archives*, I:289–93; Court of Private Land Claims, No. 225.

67   "a plaza in a square . . ."   Ibid.

67   "of a village"   Mead, *Cultural Patterns and Technical Change*, p. 169.

68   "there is not land or waters . . ."   Twitchell, *Spanish Archives*, I:290.

68   "a ragged lot . . ."   Ibid., p. 99.

68   seven racial categories   Ríos-Bustamante, "New Mexico in the Eighteenth Century"; Aragon, "The People of Santa Fe in the 1790's."

69   *indio, genízaro*   Ibid. Fray Angelico Chavez, "Genízaros" in Ortiz, *Handbook*, pp. 198–200.

69   selling as slaves   Kessell, *Kiva, Cross, and Crown*, pp. 366–67.

70   given their freedom   Chavez, "Genízaros," p. 199.

70   69 percent of the marriages   Ríos-Bustamante, p. 367.

70   simply to be *mejicano*   Chavez, "Genízaros," p. 199. To-day there exists less general agreement on the proper nomenclature for Southwestern society than there did in the 1820s and 1830s. The word *Spanish*, used broadly, suggests a connection to Spain that is more fantasy than fact, and *Spanish-American* only adds grammatical awkwardness to the inaccuracy. *Chicano*, meanwhile, connotes a political identity not universally accepted among New Mexicans, and *Mexican* tends to obscure the regional distinctiveness of New Mexican society, which Zebulon Pike noted as early as 1803. Most writers today refer to Spanish-speaking New Mexicans by the term *Hispano*, which recommends itself chiefly because it suggests an ethnic heritage and nothing more.

70   achievement in bringing water   Domínguez wrote, "This settlement is on a high level site provided by a ridge of the aforesaid sierra, with very good lands, although there is no river. But since the Almighty gave man what he needs, those interested in these lands, with prodigious labor, dammed up in a small canyon the water of a little rivulet that came through it, which arises in the east of the sierra itself. By making it rise in the dam to a height of 60 or more *varas*, they succeed in using it very freely for irrigation by means of a good ditch." *Missions*, p. 83. Domínguez's account of a dam 175 feet high is certainly an exaggeration, but the existence of a dam of any kind suggests that the upper reaches of the original *acequia madre* of Truchas followed a course different from their present one.

72   "We beg your Lordship . . ."   "*Representación hecha por los Vecinos de la nuestra Señora del Rosario de Las Truchas*," *Spanish Archives of New Mexico*, State Records Center and Archives, Santa Fe, Item 666, Roll 10, Frames 712–15.

72   shortage of firearms   Kenner, pp. 38, 48.

73   "to finish the plaza . . ."   Mendinueta, *Spanish Archives*, Item 666.

75   "it would seem difficult . . ."   A. B. Thomas, *Plains Indians*, p. 59.

CHAPTER 5

75   *"Los Comanches"*   Campa, "Los Comanches."
75   seize any Comanches   Thomas, *Plains Indians*, p. 44.
75–76   "In the evening . . ."   Ibid., pp. 175–76.

5. LITHIC PARALLELS

79   "On the side of Jicarita . . ."   New Mexico Writers Project,
        Taos County File, State Records Center and Archives.
79   ciboleros could ride out   Kenner, pp. 100–101.
80   Ojo Sarco   It is difficult to say exactly when a community
        like Ojo Sarco was established, for the process by which
        it "hived off" from Las Trampas was certainly very
        slow and gradual. At first Tramperos would have used
        the lands in the narrow Ojo Sarco valley for grazing
        and dry farming, eventually building summer huts and
        later permanent houses as pressure from Indian raiders
        permitted and population required. By the time a church
        was licensed and built, the village would probably have
        been well established for several decades. See gener-
        ally, E. Boyd, *Popular Arts of Spanish New Mexico.*
80   Santa Barbara Grant   U.S. Bureau of Land Management,
        Surveyor General Reports, Claim 114, File 122, Reel
        23. Also Domínguez, p. 92.
80   New Mexico's sheep herds   Denevan, "Livestock Num-
        bers," p. 699.
80   San Miguel del Vado   Kessell, *Kiva, Cross, and Crown,*
        pp. 416–19.
80   Lo de Mora   The date of the founding of Mora is not
        precisely known, in part because the county court-
        house and all the documents it contained were de-
        stroyed when the U.S. Army leveled Mora with artillery
        in 1847 (see Ch. 7). Fray Angelico Chavez has indicated
        that migration into the valley began about 1816 or
        1817, if not sooner ("Early Settlements in the Mora
        Valley"). Dolores Gunnerson, however, appears to have
        pushed back the date of settlement with the discovery
        of a document dated October 30, 1795 that mentioned
        a mission having been founded in a place called "lo de
        Mora" (*The Jicarilla Apaches*, p. 275).
81   prospecting expedition   Twitchell, *Leading Facts*, p. 445;
        Warner, *Domíngues-Escalante Journal*, pp. 12, 26.

81   greatest culinary delicacy   Ruxton, *Ruxton of the Rockies*, pp. 256–57.

82   great Pleistocene glaciers   The various glacial periods during the Pleistocene are identified for the southern Sangres and discussed in L. L. Ray, "Glacial Chronology of the Southern Rocky Mountains."

83   Precambrian times   Elmer H. Baltz, "Resume of Rio Grande Depression in North-Central New Mexico," in Hawley, *Guidebook to Rio Grande Rift in New Mexico and Colorado*, p. 222.

83   western third of the Pecos Wilderness   The material in the following pages which relates specifically to the geology of the Pecos high country was drawn from Miller, Montgomery, and Sutherland, *Geology of Part of the Southern Sangre de Cristo Mountains*. A good layman's introduction to the geology of the area is Sutherland and Montgomery, *Trail Guide to the Geology of the Upper Pecos*.

6. PANDORA'S SCOUTS

89   "Oh we are approaching . . ."   Gregg, *Commerce of the Prairies*, p. 101.

89   adventurous, dirty infidels   Tyler, "Anglo-American Penetration of the Southwest," p. 331. Campa, *Hispanic Culture in the Southwest*, p. 111.

89   fine furs from beaver   Weber, *Taos Trappers*, pp. 15–16, 30–31. See also Carroll, *Three New Mexico Chronicles*, p. 90.

91   isolation . . . from new ideas   Goetzmann, *Exploration and Empire*, pp. 53–55. Weber, pp. 32–33.

91   Luis María Mora   Kessell, *Kiva, Cross, and Crown*, p. 387.

92   "abounded with beaver"   Weber, p. 46.

92   pass the winter in San Miguel   Ibid., p. 54.

92   Ewing Young and William Wolfskill   Wilson, *William Wolfskill*, p. 39. Holmes, *Ewing Young*, p. 19.

93   not subject to customs duties   Carroll, p. 108. Weber, p. 57.

93   "Touse"   Ibid., pp. 53–56.

CHAPTER 6

93   one stream . . . trapped out Fowler, *Journal*, p. 138.

94   find the beaver drowned Chittenden, *The American Fur Trade of the Far West*, pp. 820–21.

94   leading Anglo explorers Hafen, *The Mountain Men and the Fur Trade of the Far West*, vols. 1, 2, 4, 5.

97   "north of Santa Fe . . ." Quoted by Holmes, *Ewing Young*, pp. 20–21.

97   already . . . trapped out Lavender, *Bent's Fort*, p. 59; Weber, p. 65.

97   harvest amounted to $10,044 Hulbert, *Southwest on the Turquoise Trail*, p. 84.

97   observed to Governor Narbona Weber, p. 82.

97   about two thousand animals The price for beaver in St. Louis seems to have stayed in this range throughout the 1820s. Ibid., pp. 100, 173. The average pelt weighed about 1.6 lbs. (ibid., p. 161).

97   no more than six thousand Carl G. Berghofer, "Protected Furbearers," in New Mexico Department of Game and Fish, *New Mexico Game Management*, pp. 187–89.

97   Storrs noted Hulbert, p. 93.

98   first conservation measures Weber, pp. 224–25.

98   infusion of capital and . . . know-how Carroll, p. 39; Weber, p. 104.

99   buying, selling . . . pretty much at will. Ibid., pp. 66–67, 105 ff., 156, 176 ff.

99   Narbona complained Ibid., p. 115.

99   never passed through Santa Fe Moorhead, *New Mexico's Royal Road*, pp. 63–65; Carroll, p. 66.

100   "The *patrona* of the family . . ." Ruxton, *Ruxton of the Rockies*, p. 188.

100   not a printing press Gregg, p. 142. Carroll, XX, 36–38, 95–96.

100   "crafts are in the worst . . ." Ibid., p. 38.

100   "There are no people . . ." Weber, *Foreigners in Their Native Land*, p. 72.

101   two dozen clergymen Carroll, p. 50. Hulbert, p. 76.

101   cruel and cowardly Paredes, "The Mexican Image in American Travel Literature."

101   "The human race . . ." Parkman, *The Oregon Trail*, p. 332.

7. Discord and Dollars

103   "Do not find it strange ..." As quoted by Weber, *Foreigners in Their Native Land*, pp. 128–29.

105   Kearny assured the local populace Twitchell, *Leading Facts*, 2:206–7.

105   "for secret services rendered" Benton, *Thirty Years View*, 2:684.

105   "was of a different mould ..." Ibid., p. 623.

106   Charles Bent ... scalped Exec. Doc. No. 8, 30th Cong., 1st Sess., pp. 520–38. These official reports of military actions during the insurrection in New Mexico are reproduced and supplemented in Senate Doc. 442, 56th Cong., 1st Sess., pp. 1–33. The exact number of Americans killed at Mora remains unclear. Early dispatches reported four, but in subsequent accounts the number varied as high as eight.

108   "If we had one or two pieces ..." Ibid., p. 20.

109   "that sink of vice ..." As quoted by Emmet, *Fort Union and the Winning of the Southwest*, p. 12.

109   Fort Union's military mission Utley, *Fort Union*, pp. 9, 50–52.

111   St. Vrain Hanosh, "A History of Mora." W. S. Wallace, *A Journey Through New Mexico's First Judicial District*, pp. 46, 55.

111   Watrous Perrigo, "The Original Las Vegas," p. 92.

111   population grew Zeleny, *Relations between the Spanish-Americans and Anglo-Americans in New Mexico*, pp. 126–27. The figure of 3,000 thousand Anglos in New Mexico in 1860 was obtained by adding the number of persons counted in the 1860 census as born in the U.S. exclusive of New Mexico (1,168) to the number of foreign born persons who were not native of Mexico (1,908).

112   "The rapid and irresistible ..." W. S. Wallace, p. 43.

112   the Jicarilla in 1854 Taylor, "Campaigns Against the Jicarilla Apache, 1854," pp. 274–75.

113   "were driven into the war ..." As quoted by Emmet, p. 177.

114   a succession of ... campaigns Utley, pp. 19–22, 37–45.

114   no question about ... loyalty Kenner, pp. 90–92. Lav-

ender, *Bent's Fort*, Ch. 18. Calhoun to Brown, 25 January 1850, in Abel, *Official Correspondence of James S. Calhoun*, pp. 104–5.

116   aided their Indian friends   Kenner, pp. 148–49.

116   assisted Comanches   Ibid., pp. 166–67.

116   entrepreneurs purchased stolen herds   Emmet, p. 358; Kenner, p. 173.

116   others were Hispanos   Emmet, p. 358; Kenner, pp. 170–72.

117   "days filled with feasting . . ."   Brown, *Hispano Folklife of New Mexico*, p. 49.

117   specifically mentions blankets   Ibid., p. 44.

118   traded human beings   Kenner, pp. 94–96.

118   "This was a good offer . . ."   Brown, p. 47.

119   "many escapes . . ."   Ibid., p. 48.

119   "Salvo to San Antonio"   Ibid., p. 49.

8. DISCIPLINE

121   "Adios, friends . . ."   Translation by author. The original Spanish is reproduced in Marta Weigle, *Brothers of Light, Brothers of Blood: The Penitentes of the Southwest*, p. 191:

> Adios por ultima vez
> Que me ven sobre la tierra,
> Me echan in la sepultura
> Que es la casa verdadera.

121   world of Vicente Romero   Dickey, *New Mexico Village Arts*, pp. 38, 51, 101. The villagers began to yoke their oxen at the shoulders only after they learned to do so from Anglo traders on the Santa Fe Trail. Up to that time they had simply attached the beam of their plow directly to a stout pole lashed to the oxen's horns; all in all, a very inefficient way to harness animal power.

122   regional brand of Spanish   Domínguez, p. 42. Also, Campa, *Hispanic Culture*, pp. 217–25.

123   Historians disagree   Chavez contends that the Brotherhood was fully developed elsewhere in the Spanish New World and imported to New Mexico sometime between 1790 and 1810. See Chavez, "The Penitentes of New Mexico," pp. 97–123. Weigle, on the other hand,

argues persuasively that the Brotherhood was an out-
growth of the Third Order of Saint Francis, whose history in New Mexico goes back at least as far as the
Reconquest. See Weigle, *Brothers of Light*, pp. 26–51.

123   strong and ritually complete   Weigle, pp. 24–25, 51.

123   undermanned and neglectful   Carroll, *Three Chronicles*,
pp. 50–55.

124   "Spiritual administration . . ."   Ibid., p. 53.

124   population almost trebled   Zeleny, *Relations between the
Spanish-Americans and Anglo-Americans*, pp. 26–27,
and Meinig, *Southwest*, pp. 27–35.

124   morada might be built   Boyd, *Popular Arts of Spanish New
Mexico*, pp. 459–60; Bunting, "An Architectural Guide
to Northern New Mexico," pp. 13–51. Both Trampas
and Truchas used to have moradas located some distance from the village center.

125   surrogate Christs   Weigle, pp. 171–73.

126   fatal human hit   Ibid., pp. 168–70.

127   fraternity and community   Ibid., p. 139.

127   reinforce the cultural pride   Ibid., pp. 150–53; González,
The Spanish-Americans of New Mexico, pp. 86–88.

128   Hispanos had all the vices   Paredes, "The Mexican Image
in American Travel Literature," pp. 23–25.

128   general stores sprang up   For a detailed examination of
the "general store" trade and the growth of mercantile
capitalism in northern New Mexico see William A.
Parish, *The Charles Ilfeld Company*.

128   "deliberate, organized attempts . . ."   Wallace, *Culture and
Personality*, p. 188.

129   Handsome Lake   See Wallace, *The Death and Rebirth of
the Seneca*.

129   a majority of Hispanic males   Weigle, p. 96.

129   "auxiliadores"   Ibid., pp. 144–45.

130   delay of . . . statehood   Ibid., p. 79.

130   Jean Baptiste Lamy   Horgan, *Lamy*, pp. 283, 314–20, and
passim.

130   Penitentes resisted   Weigle, pp. 52–63, 75–76.

131   Chaves, who acted as bodyguard   Simmons, *Two Southwesterners*.

132   increasing worldliness   Weigle, pp. 96–101.

132   endorsement of Catholic Church   Ibid., p. 111.

CHAPTER 9

132   membership of . . . less than two thousand   Ibid., p. 117.
132   loss of social stability   Knowlton, "Changing Spanish American Villages," p. 471.

9. THE VIEW FROM THE TOP OF THE CLIFF

135   Giovanni Maria Agostini   The most detailed treatment of the hermit is Campa, *Treasure of the Sangre de Cristos*, pp. 161–96. See also Omar and Elsa Barker, "Hermit of the Mountain"; Lynn I. Perrigo, "The Original Las Vegas," pp. 109–11; and Phil LeNoir, "The Hermit of Las Vegas."

136   too independent to submit   Cf. Campa, p. 190.
136   El Cerro del Tecolote   Ibid., p. 185.
137   sheaf of passports   The main source of biographical information on the hermit is Charles Wolfe, "The History of the 'Wonder of Our Century,'" a documentary biography of Agostini that Wolfe never completed. It consists of what Wolfe claims in his "Introduction" are faithful translations of Agostini's personal papers, to which Wolfe had access, probably in the 1930s or earlier. Campa describes these materials and his own attempts to locate the originals in *Treasure*, Ch. 12.

138   killed his cousin   Perrigo, "Las Vegas," p. 109.
140   110 deaths from smallpox   Hanosh, "A History of Mora."
140   dagger in its back   Campa, *Treasure*, p. 194.
141   La Sociedad del Ermitaño   Ibid., pp. 166, 189.
141   El Porvenir   Perrigo, "Las Vegas," p. 433.
141   annual pilgrimage   Ibid., p. 434.
144   a signal fire   Campa, *Treasure*, pp. 190–91; Barker, "Hermit," p. 355.

10. GOLD IN THEM HILLS

147   "That Hermit . . ."   Edwords, *Campfires of a Naturalist*, p. 84.
149   rushed to Old Baldy   Twitchell, *Leading Facts*, 3:66; Jenkinson, *Ghost Towns of New Mexico*, p. 16.
149   lawlessness . . . federal troops   Ibid., pp. 17, 20.
149   four million dollars   Christiansen, *The Story of Mining in New Mexico*, pp. 37–38.

149   "but indifferently prospected"   New Mexico Bureau of Immigration, *The Mines of New Mexico*, pp. 1–5.

150   busiest and richest town   Perrigo, "Las Vegas," p. 132.

150   built a cabin   Edwords, pp. 80 ff.; E. S. Barker, *Beatty's Cabin*, pp. 1–8.

151   gold . . . south from Elk Mountain   Perrigo, "Las Vegas," p. 170.

151   hauled up into the Gallinas hills   Ibid., p. 171.

151   Rociada ores   Ibid. See also F. Jones, *Old Mining Camps of New Mexico*, p. 77.

152   preposterous stories   Edwords, pp. 80 ff.

154   deserted Pecos cabin   E. Barker, p. 2.

155   the Mosimanns took it   Harry Mosimann, Las Vegas, New Mexico, January 1977, personal communication.

155   named Evangeline   New Mexico Geological Society, *Guidebook*, p. 140.

155   American Metal Company   Ibid.

155   New Mexico's leading producer   Sherman, *Ghost Towns and Mining Camps*, pp. 206–7. Also, Henderson, *Gold, Silver*, pp. 531, 553–54.

155   boom years   Geological Society, pp. 140–41; Sherman, *Ghost Towns*, p. 207.

156   Terrero flourished   Ibid.

157   underground flooding   Geological Society, p. 141.

157   became a ghost town   Sherman, p. 207.

157   major oil companies   Planning Staff, Supervisor's Office, Santa Fe National Forest, August 1981, personal communication.

158   "the whole channel . . ."   F. Jones, *Old Mining Camps*, pp. 79–80.

159   Jealous Frenchmen   Applegate, *Native Tales of New Mexico*, pp. 181–91.

159   clerk at Beaubien's store   Weber, *Taos Trappers*, pp. 219–20.

11. INVENTORY

163   "To a great extent, . . ."   J. W. Powell, *Report on the Lands of the Arid Region of the United States*, p. viii.

164   outlaw Jicarilla   A. C. Greene, *The Last Captive*, pp. 83–85.

164 "Geographical Surveys" Wheeler, *Annual Report upon the Geographical Surveys West of the One Hundredth Meridan*, pp. 43–44.

165 restore . . . the luster Bartlett, *Great Surveys of the American West*, pp. 336–38.

165 "astronomical, geodetic . . ." Ibid., p. 338.

165 the causes of science Goetzmann, *Exploration and Empire*, p. 482.

165 impressive scientific coups Ibid.

165 investigations of the Pecos Wheeler, *Annual Reports*, 1875–77.

168 "there are indications . . ." Ibid., 1876, pp. 45–46.

168 "with all its . . . advantages . . ." Ibid., p. 129.

168 "In no part . . ." Ibid., p. 128.

168 "of supporting ten times . . ." Ibid.

169 "The writer has visited . . ." Ibid., p. 129. The Pecos River is not a tributary of the Canadian.

169 "a peculiar state . . ." Ibid.

169 "It would be economy . . ." Ibid., pp. 129–30.

169 play a manager's role Goetzmann, *Exploration and Empire*, p. 487, n. 4.

170 abolished the Wheeler Survey Ibid., p. 487. Bartlett, pp. xv, 213, 374.

170 corruption and lawlessness Lamar, *The Far Southwest*, pp. 151, 169, and generally Chs. 6 and 7.

12. FRACTIONS OF JUSTICE

171 "I have just been offered . . ." Bond to Warshauer, 20 February 1903, University of New Mexico, General Library Special Collections (UNM-SC), Archive 133, Item 76, p. 21.

171 began his tenure Westphall, *The Public Domain in New Mexico*, p. xiii.

171 economy . . . roll forward Lamar, *Far Southwest*, pp. 139, 149.

172 yield a fortune Westphall, *Public Domain*, p. 51.

174 legal limbo Ibid., pp. 3–4, 49–52.

174 one-third . . . interest Ibid., p. 49.

175 selling his land White et al., *Land Title Study*, p. 32.

175   scattered along the edges   Simmons, "Settlement Plans and Village Patterns," p. 17.

177   ejido . . . not legally be sold   Malcolm Ebright, *The Tierra Amarilla Grant: A History of Chicanery*, p. 5, n. 14, and personal communication 19 March 1982. Many writers on New Mexican history, including the present one, have wrongly argued that the reason the ejido might not be legally sold was that it remained the property of the sovereign who held it in trust for the settlers. See, William deBuys, "Fractions of Justice: A Legal and Social History of the Las Trampas Grant, New Mexico," p. 74. Ebright discusses this tradition of historical error in "The San Joaquín Grant: Who Owned the Common Lands? A Historical-Legal Puzzle."

178   Congress confirm   U.S. Bureau of Land Management, *Surveyor General Reports*, Claim No. 65, File No. 27.

178   four successive surveys   Ibid.

178   David Martínez, Jr.   State Records Center and Archives (SRCA), Santa Fe District Court Records, Cause 3369.

179   largest single interest   Martínez to Surveyor General Hobart, 26 November 1892, *Surveyor General Records*, Claim 65, File 27.

179   legislation to inhibit   Keleher, "Law of the New Mexican Land Grant," p. 364.

179   "and if partition . . ."   SRCA, Taos Court, District Court Records, Cause 594. Almost all the material having to do with this case, the Suit for Partition is duplicated in the file of Cause 840, the suit to quiet title brought by the Las Trampas Lumber Company against Ortega et al. in 1908.

180   appointment of a referee   "Order of Reference," 22 February 1901 (as duplicated in an untitled, unsigned document in file for Cause 840, hereinafter referred to as "Brief").

180   Amado Chaves . . . silent partner   SRCA, Santa Fe District Court Records, Cause 4329.

180   a mere 650 acres   Cause 840, "Brief," p. 6.

180   Johnson discovered   Ibid., pp. 6–8.

181   "either by consanguinity . . ."   Ibid., p. 9.

181   Amado Chaves's brother   Cause 840, "Answer to Complaint," p. 12.

182   "owing to the character . . ."   Ibid., p. 10.

182   "be sold at public auction   Ibid., "Exhibit B," a copy of the decree issues in Cause 594.

182   Baca group filed suit   SRCA, Santa Fe District Court Records, Cause 4399.

182   closed the sale   Cause 840, "Brief," pp. 11–12.

183   canceled the sale   Ibid., pp. 13, 15.

183   lucky to net $380   Ibid., pp. 11–12. Costs were as follows:

Survey and platting	$ 445.39
Johnson's services as referee	190.00
" " " Special Master	500.00
Expenses reimbursed to McMillen	62.30
Commissioners' fees ($5 each)	15.00
Taos County taxes, 1893–1902	
(reduced from $4,543)	800.00
Rio Arriba County taxes (est. minimum)	200.00
Total Costs:	$2,212.69
Balance from $5,000:	2,787.31
Share of David Martinez (18.3%):	$510.07
Balance after deduction of	
McMillen's fee (25%):	$382.56

184   Rio Arriba . . . served notice   SRCA, Docket Book of Rio Arriba County for 1902.

184   McFie named the bank   Cause 840, "Brief," pp. 15–16.

184   proceeded with the auction   Ibid., pp. 16–17.

184    $17,012   Ibid.

185   "undoubtedly the best . . ."   F. Bond to G. Bond, 25 March 1903, UNM-SC, Archive 133, Item 76, p. 35.

185   "My idea is . . ."   Ibid., p. 34.

186   "Young Catron . . ."   Ibid.

186   "It might be that . . ."   Ibid. Bernard S. Rodey was an influential Albuquerque lawyer.

186   Bond wrote to the commissioner   Ibid., p. 37.

186   make a down payment   "Articles of Incorporation of the Las Trampas Lumber Company," filed 11 June 1907, Bernalillo County. U.S. Department of Agriculture, Forest Service, Region 3 Headquarters, Albuquerque, copy kept with miscellaneous land title records pertaining to the Las Trampas Land Grant.

187   journey to Santa Fe   Interview with Jacobo Romero, El Valle, New Mexico. Romero's father, Narciso Romero, was one of the men making the trip with Ortega.

187   stated their case   Cause 840, "Answer to Complaint."

188   Bond . . . assumed . . . responsibility   Frank H. Grubbs, "Frank Bond: Gentleman Sheepherder," p. 314.

188   worked out an agreement   Cause 840, "Agreement" and "Judgment and Decree," p. 475, SRCA, Book D District Court Records, Eighth Judicial District.

189   excluded from partition   Cause 840, "Answer to Complaint," p. 18.

190   grant up for sale   Bond to J. B. Herndon, President of the Las Trampas Lumber Company, 10 March 1914, UNM-SC, Archive 133, Item 78, p. 415. Also, Cause 840, "Advertising Memorandum."

190   "A map of the Trampas . . ."   Bond to J. B. Herndon, 1 April 1914, Archive 133, Item 78, p. 583. Bond's public statements contrast markedly with this private one. See Frank Bond, "Memoirs of Forty Years in New Mexico."

190   Breece Lumber Company   Forest Service, Region 3 Headquarters, miscellaneous land title records pertaining to the Las Trampas Grant.

190   Cañon de San Diego   U.S. Dept. of Agriculture, Soil Conservation Service, "Notes on Community-owned Land Grants in New Mexico."

191   kept the rest of the money   Interviews with Jacobo Romero and Juan de Dios Romero, August 1979.

192   Gorras Blancas   Rosenbaum, "Mexicano vs. Americano."

192   prohibited Pueblos   Simmons, "History of the Pueblos Since 1821," in Ortiz, *Handbook*, pp. 214–15.

13. MANITOS

193   "Mountain villages: . . ."   Shiki, "Winter," in Henderson, *An Introduction to Haiku*, p. 180.

193   Hispanic villages of the Sangres   Unless otherwise noted information in this chapter concerning village life and customs was obtained in conversations with Jacobo Romero and Eloisa Romero and other residents of the village of El Valle during the period 1975–81.

195   "People now . . ."   Interview with Amado Aragón, Mora,

June 1975. Mr. Aragón grew up in the village of LeDoux in the early part of this century.

195   *verguenza*   Paul Kutsche, editor, *The Survival of Spanish American Villages*, pp. 18, 99–104.

196   agricultural revolution   See, for instance, Bidwell, "The Agricultural Revolution in New England."

199   deep reverence for the land   Swadesh, *Primeros Pobladores*, p. 195.

203   herbs were credited   Curtin, *Healing Herbs of the Upper Rio Grande.*

203   typhoid   Weigle, *Hispanic Villages of Northern New Mexico*, passim.

204   *partidario*   Ibid., pp. 213–22.

204   Bond and Sargent   Swadesh, *Primeros Pobladores*, pp. 115–16. Frank H. Grubbs, "Frank Bond: Gentleman Sheepherder," 35:199; 37:67–68. The Forest Service was not necessarily a willing accomplice in the extension of Bond's and Sargent's commercial empire. The Forest Service generally did what it could to make grazing permits available to small ranchers, but many of those ranchers, submitting to a variety of pressures, sold or transferred their permits to the Anglo entrepreneurs (see Ch. 15).

205   slid deeper in debt   Frequently, but not always. William J. Parish argues that the partido system, as employed by the Charles Ilfeld Company of Las Vegas, was in many cases beneficial to the *partidarios*. Parish, *The Charles Ilfeld Company*, pp. 150–73.

206   ample supply of cheap ... labor   Weigle, *Hispanic Villages*, pp. 213–22.

206   one member of every ... family   Ibid., p. 224.

207   average holding of ... land   Carlson, "The Rio Arriba: A Geographic Appraisal of the Spanish-American Homeland," pp. 81–82.

207   becoming businessmen   Ann Mayhew, "A Reappraisal of the Causes of Farm Protest in the United States, 1870–1900."

207   communities of the Rio Abajo   John R. Van Ness, "Hispanic Village Organization in Northern New Mexico," in Kutsche, *Survival of Villages*, pp. 23–44.

207   cash penetrated   Cf. Andrew Pearse, "Metropolis and the

Peasant: The Expansion of the Urban-Industrial Complex and the Changing Rural Structure," in Shanin, ed., *Peasants and Peasant Societies*, pp. 69–80.

210   flow of dollars . . . slowed   Weigle, *Hispanic Villages*, pp. 224–25.

210   seriously undernourished   Ibid., p. 211.

210   on relief   Ibid., p. 36.

211   lived in . . . poverty   Bureau of Census, *1970 Census of Population*, vol. 1, part 33, pp. 33–217.

211   vast numbers of young people   The following table offers a statistical portrait of outmigration from northern New Mexico.

Rural Population Change in Five Northern New Mexico Counties
1930–1970

| | | | | | | Percent | |
| | | | | | | Change | Inferred Migration |
County	1930	1940	1950	1960	1970	1940–70	1940–70
Mora	10322	10981	8720	6028	4673	−57.4	−69.9
Rio Arriba	21381	25352	24997	24193	21268	−0.7	−41.3
San Miguel	14539	15548	12749	9650	8116	−47.8	−49.1
Santa Fe	8391	10501	10155	10294	11963	+13.9	−4.5
Taos	14394	18528	17145	15934	17516	−5.5	−42.6

N.B.:   1. This table shows rural population only and therefore excludes the population of the cities of Las Vegas and Santa Fe from figures for San Miguel and Santa Fe counties, respectively. The populations of Taos and Española, however, are included in Taos and Rio Arriba county figures.

     2. "Inferred migration" balances net population change against birth and death rates to reflect the actual movement of people to or from an area. Urban populations are included in these figures.

Sources: (for population figures) U.S. Bureau of the Census, Sixteenth, Seventeenth, Eighteenth, and Nineteenth Census of the United States (Washington: Bureau of the Census, 1942–1972).
(for inferred migration) Kenneth R. Weber, "Rural Hispanic Village Viability from an Economic and Historic Perspective," in Kutsche, *The Survival of Spanish American Villages*, p. 87.

14. WASHED AND WORN

215   "Don't get excited . . ." As remembered by Perl Charles, forest ranger for the Upper Pecos District during the early and middle 1920s. Bond made the remark as he paid a $50 fine for range trespass by one of his herders. Charles said Bond's comment made him "a little sore." Edwin A. Tucker and George Fitzpatrick, interview transcripts for *Men Who Matched the Mountains* (1972).

216   wettest fifteen years Lee Wilson and Associates, "Future Water Issues, Taos County, New Mexico." See also Kelley, *The Contemporary Ecology of Arroyo Hondo, New Mexico*, pp. 29–33.

216   "many acres . . ." Pancoast, *A Quaker Forty-Niner*, p. 213.

216   "I had to walk . . ." Ibid., p. 216.

217   dry gravel bed Cooperrider and Hendricks, "Soil Erosion and Stream Flow on Range and Forest Lands of the Upper Rio Grande Watershed in Relation to Land Resources and Human Welfare."

217   a billion cubic meters Edward J. Dortignac, "Rio Puerco: Abused Basin," in Hodge, *Aridity and Man*, pp. 507–16. See also David Sheridan, *Desertification of the United States*.

217   erosion was partly geologic Luna Leopold, among others, has argued that the desertification of the Puerco basin was primarily a geologic and climatic event in which man played only an incidental role. The cornerstone of his argument is the fact that in at least one specific location (near Cabezon) the channel of the Rio Puerco was already deeply incised in 1847—before extensive human settlement (L. Leopold, "Vegetation of Southwestern Watersheds in the Nineteenth Century"). Leopold's influential discussion, however, does not take into account the short-lived but nonetheless intensive Hispanic settlement of the valley in the second half of the eighteenth century. (Cf. Larry López, "The Founding of San Francisco on the Rio Puerco: A Document.")

218   "In recent years . . ." Ligon, *Wild Life of New Mexico*, p. 35. See also Elmer Otis Wooten, "The Range Problem in New Mexico."

218   grasslands to decline by . . . 50 percent   U.S. Department of Agriculture—Forest Service, "The Western Range." The estimate of a 50 percent loss in productivity is an average for all of the western states. The loss was considerably greater in the Southwest, where the warmer climate permitted year-round grazing in many areas. Ferdinand E. Mollin, secretary of the American National Livestock Association, rebutted the Forest Service's analysis with *If and When It Rains: The Stockman's View of the Range Question.*

219   cattle industry . . . had grown   Denevan, "Livestock Numbers in Nineteenth-Century New Mexico and the Problem of Gullying in the Southwest." See also Haskett, "Early History of the Cattle Industry in Arizona" and Wagoner, "Overstocking of the Ranges in Southern Arizona During the 1870s and 1880s."

219   they cut steps   Interview with Amadeo Aragón, Mora, June 1975.

220   sheep population   Stocking figures vary considerably from author to author. I have relied on Carlson, "New Mexico's Sheep Industry."

222   climax grasses   The concept of climax vegetation is frequently misunderstood to signify an ecosystem that is somehow ideal because of its supposed immutability under "natural" conditions, i.e., those unaffected by man. In actual fact, all ecosystems are dynamic and not static, and few if any can be considered unaffected by human activity. By using the term *climax*, I have meant only to identify plants that dominated the mountain vegetation prior to the introduction of large herds of domestic animals.

222   *Monthly Bulletin* . . . reported   As noted by Parish, *The Charles Ilfeld Company*, p. 178.

223   greater than average rainfall   Denevan, "Livestock Numbers," pp. 700–730.

223   flooding thundershowers   Ibid. See also, Cooke and Reeves, *Arroyos and Environmental Change*, pp. 2–4, 187–89.

224   commonest features   Cooke and Reeves, *Arroyos*, pp. 2–4. Malde and Scott, "Observations of Contemporary Arroyo Cutting Near Santa Fe, New Mexico, U.S.A.," p. 39.

224   sagebrush now dominates   Gross and Dick-Peddie, "A Map of Primeval Vegetation in New Mexico"; Bohrer, "The Prehistoric and Historic Role of the Cool-Season Grasses in the Southwest." Vegetation changes in southern New Mexico were equally drastic due to invasion of grasslands by mesquite and creosotebush. See David R. Harris, "Recent Plant Invasions in the Arid and Semi-Arid Southwest of the United States," in Detwyler, ed., *Man's Impact on Environment*, pp. 459–75.

225   limits stockraising   Various interviews with range management personnel of the Santa Fe and Carson National Forests, 1975–1980.

225   Sheep fescue   Ibid. Also, Hitchcock, Manual of the Grasses of the United States; William A. Weber, *Rocky Mountain Flora*.

225   Kentucky bluegrass   Ibid. See also Carrier and Bort, "The History of Kentucky Bluegrass and White Clover in the United States; Scherry, "The Migration of a Plant."

226   "an area where the earth . . ."   Public Law 88-577, September 3, 1964, *United States Statutes at Large 78* (1965):891.

227   AT&SF . . . consumed the timber   Gjevre, *Chili Line: The Narrow Rail Trail to Santa Fe*, pp. 45–46.

227   McGaffey also purchased   Ibid. Also, "Operation Crossties," *New Mexico Magazine*, May 1954.

228   steam-powered sawmill   Gjevre, *Chili Line*, pp. 45–46.

228      $5 or $6 worth   Ibid. Wages varied over the time that Santa Barbara Tie and Pole was active, and many of the older men in the Peñasco valley remember rates slightly different from these.

228   creosoting plant in Albuquerque   Tucker and Fitzpatrick, *Men Who Matched the Mountains*, p. 124.

229   the change in flow   The principal climatic component in the changed rate of flow is a trend since the beginning of the century for a greater and greater portion of annual precipitation to fall during the summer, leaving proportionately less winter snowpack. Since annual precipitation was somewhat greater during the years of the Santa Barbara Tie and Pole operation, modern snowpacks may indeed be puny compared to those of the 1910s. Nevertheless, it appears clear that changes

in forest cover also played a major role. Wilson, "Water Issues, Taos County," Ch. 4; Peter Stewart, Hydrologist, Carson National Forest, "Watershed Characteristics Part B: Decreases in Flow in the Rio Hondo," March 1981.

230 transpired by plants P. Stewart, "Watershed Characteristics."

230 cultivation ... abandoned Cooperrider and Hendricks, "Soil Erosion and Stream Flow," pp. 14–20, 85–87.

231 disastrous proportions Harper et al., *Man and Resources in the Middle Rio Grande Valley*, pp. 33–47.

231 many ranchers ... lost everything Roberts, *Hoof Prints on Forest Ranges*, pp. 121–22; William Voigt, Jr., *Public Grazing Lands*, pp. 53–57.

232 serious, accelerated erosion Cooperrider and Hendricks, pp. 2–3.

232 fed cactus Weigle, *Hispanic Villages*, pp. 116, 200..

232 villagers lived in harmony See, for example, Peter van Dresser, *A Landscape for Humans*.

232 guardians of a careful balance Cf. John L. Kessell, "Spaniards, Environment, and the Pepsi Generation," in Weber, ed., *New Spain's Far Northern Frontier*, pp. 287–91.

233 overgrazing and deforestation The classic treatment of this subject is George Perkins March, *Man and Nature*. See also Eckholm, *Losing Ground*.

233 fourteen million acres United States, *United Nations Conference on Desertification*.

15. BULLY BOYS AND BUREAUCRATS

235 "I hand you ..." R. C. McClure to Postmaster General, 1 March 1901, typed copy among interview transcripts for Tucker and Fitzpatrick, *Men Who Matched the Mountains*, p. 51.

235 California's ... lesson Burcham, "Cattle and Range Forage in California, 1770–1880"; Pulling, "Range Forage and California's Range Cattle Industry."

236 self-restraint was self-punishment See Hardin, "The Tragedy of the Commons."

236	fares for the Northern Pacific Sax, *Mountains Without Handrails: Reflections on the National Parks*, p. 6.
237	undebated amendment Pinchot, *Breaking New Ground*, p. 85.
237	voted no funds Hays, *Conservation and the Gospel of Efficiency*, p. 36.
237	Forest Management Act Ibid., pp. 36–37.
238	first inspection tour Elliot S. Barker, *Beatty's Cabin*, pp. 38–39.
238	"blatant Bryan Free Silver Democrat . . ." McClure to Postmaster General, 1 March 1901; also McClure to C. M. Foraker, U.S. Marshall, 11 January 1901; William Sparks to L. E. Osenton, Cleveland District Ranger, 18 January 1901; and D. C. Kerley, Pecos District Ranger, to Commissioner of the General Land Office, 28 February 1901, interview transcripts for Tucker, *Men Who Matched the Mountains*, pp. 50 ff.
239	change the preservationist image Hays, *Conservation*, pp. 38–48.
239	join the Klondike gold rush Interview with Elliot S. Barker, 6 June 1977, Santa Fe.
240	sixty dollars a month Interview transcripts for Tucker, *Man Who Matched the Mountains*.
241	arsonists had set fire Ibid., R. C. McClure to D. C. Kerley, 23 May 1901, p. 70.
241	miles from the nearest herders Tucker, *Men Who Matched the Mountains*, pp. 34–36.
242	"Sure enough . . ." Ibid.
242	"there were twice . . ." John W. Johnson, *Reminiscences of a Forest Ranger, 1914–1944*, p. 29.
245	"Mr. Johnson . . ." Ibid.
245	elimination of feral horses Ibid., pp. 42–43.
245	"the grazing lands . . ." Swadesh, Vigil, and Ochoa, "The Lands of New Mexico," p. 33.
246	Bond and Sargent Interview with Fred Swetnam, Española District Ranger, 10 June 1980, Española. Over a long career Mr. Swetnam also served as District Ranger at Peñasco, Jemez Springs, Las Vegas, and other districts. Also: Elliot Barker, "The Perils of Range Management," in Myers, *The Westerners*, pp. 58–59.

246   large . . . producers do not lease   Interview with M. Jean Hassell, Regional Forester and William Snyder, Director of Range Management, Albuquerque, 11 April 1984.

246   limits the maximum size   Ibid.

246   fewer than 50   U.S. Dept. of Agriculture—Forest Service, 1980, "Analysis of the Management Situation," Santa Fe National Forest Land Management Planning, and 1974, "Analysis of Grazing Situation," Carson National Forest.

247   policy . . . to favor . . . small operators   Roberts, *Hoofprints*, pp. 45–46.

247   abandonment of sheep and goat ranching   Interviews with M. Jean Hassell, Phil Smith, Fred Swetnam, U.S. Forest Service, Region 3, various dates.

248   2,200 individuals held permits   William Snyder, personal communications, 19 September 1984.

249   twenty-two or fewer cattle transferred   Interview with Hassell and Snyder, 11 April 1984.

249   Buckskin Manual   Steen, *The U.S. Forest Service: A History*, p. 78.

250   *esprit de corps*   Robinson, *The Forest Service: A Study in Public Land Management*, p. 261.

251   staff and appropriations increased   Steen, pp. 280–90.

251   "The object of . . . policy . . ."   Pinchot to Robert U. Johnson, 27 March 1904, as quoted by Hays, p. 42.

251   earned the enmity   Robinson, pp. 14–16; Shepherd, *The Forest Killers*, p. 34.

252   environmental legislation   See, U.S. Dept. of Agriculture, *The Principal Laws Relating to Forest Service Activities*, pp. 197–98, 201–8, 343–59.

252   Elk Mountain Road   Shepherd, *Forest Killers*, pp. 183–213.

253   Ski Area . . . borrow land   Ibid., p. 191. Shepherd cites a letter from Ernie Blake, director of the Taos Ski Valley, to Governor David Cargo, pointing out the inadequacies of Elk Mountain as a ski resort.

16. LAND AND CATTLE

255   "Most of all . . ."   William Hurst, Regional Forester to

CHAPTER 16

Deputy Regional Foresters and Forest Supervisors, c. 1 June 1968, memorandum.

257 consists of patented ... grants Hurst to Forest Supervisors and District Rangers, 6 March 1972, memorandum.

258 Reies López Tijerina Among the sources treating the career of Tijerina and the Alianza are Nabokov, *Tijerina and the Courthouse Raid*; Steiner, *La Raza*; Knowlton, "Violence in New Mexico"; and Swadesh, "The Alianza Movement of New Mexico," in Tobias and Woodhouse, *Minorities and Politics*.

258 grievances concerning ... grazing Alianza Federal de las Mercedes, "Spanish Land Grant Question Examined" reprinted in Cortes, *Spanish and Mexican Land Grants*.

259 cut by 60 percent File records for the Sierra Mosca grazing allotment, Supervisor's Office, Santa Fe National Forest.

259 Canjilon ... lost permits Steiner, pp. 6–7.

259 only 14,400 and 25,200 Jean M. Hassell, "The People of Northern New Mexico and the National Forests," p. 13.

260 Alianza argued Swadesh, "The Alianza Movement," pp. 75–76.

261 serious civil violence The material presented relating to the courthouse raid, manhunt, murder of Eulogio Salazar, and trial is from Nabokov, *Tijerina*, pp. 70–116, 226–68, and passim.

263 A state representative ... complained Ozarks—Four Corners Regional Development Commission, *Hearings*, pp. 520–22.

264 "make an even stronger effort ..." Freeman to Edward P. Cliff, Chief, U.S. Forest Service, 28 October 1967. Copy of letter obtained from USFS Regional Office, Albuquerque.

265 doubling of budget allocations Larry R. Coffelt, Director of Program Planning and Budget, Forest Service Regional Office, Albuquerque: personal communication, 2 September 1980.

265 "behind the rest ..." Hassell, "People of Northern New Mexico," p. 2.

265   recommended . . . measures, . . . relating . . . to grazing   Ibid.,
pp. 12–23.

266   trespass fines   Ibid., p. 16.

266   "The unique value . . ."   Hurst to Forest Supervisors and
District Rangers, memorandum titled "Region 3 Policy on Managing National Forest Land in Northern
New Mexico," 6 March 1972.

267   total grazing obligation   "Analysis of the Management
Situation," Santa Fe National Forest, 1980.

268   created new rangelands   Interviews with Wendell Gore,
Range Staff, Santa Fe National Forest, 11 July 1980,
and Pete Tatschl, Pecos District Range Conservationist, 27 June 1980.

268     $2.5 million on . . . improvements   Coffelt, personal communication, 2 September 1980. Over the same ten-year period the Carson National Forest spent $2.4 million on range administration.

269   calculate . . . benefits more liberally   Hassell, "People of
Northern New Mexico," p. 14, and interviews with
Phil Smith and Ray Dalen, Range Staff, Forest Service
Regional Office, Albuquerque, July 1980.

269   ranchers themselves also subsidize   Gray, "Economic
Benefits from Small Livestock Ranches in North-Central New Mexico."

269   obtain a grazing permit   Eastman and Gray, "Social Analysis Model for the Santa Fe National Forest," p. 16.

270   They persist in ranching   These views were expressed by
all of the thirteen Santa Fe National Forest grazing
permittees I interviewed during the summer of 1980
while making a study of range management practices
in northern New Mexico for the John Muir Institute
for Environmental Studies of Napa, California. In every
interview the permittee was guaranteed anonymity.

271   Santa Fe National Forest was overstocked   Land Management Planning data, Supervisor's Office, Santa Fe National Forest, June 1980.

272   "I could not . . ."   Interview with John Donald, former
range conservationist, Penasco District, Carson National Forest, August 1980. Cf. discussion of "The Principle of Conservation of Cognitive Structure" in
Anthony F. C. Wallace, *Culture and Personality*.

Chapter 17

272   "We are natives . . ."   Interview with a Santa Fe National
       Forest permittee, July 1980.

273   destroyed the pastoral base   See, for instance, John Ni-
       chols, *If Mountains Die*, and Stan Steiner, *La Raza*.

274   problems of the Third World   See Erik Eckholm, *Losing
       Ground: Environmental Stress and World Food Pros-
       pects*.

276   authorized a harvest . . . of over 1,700 cords   As one of
       the concerned local residents, I myself made the anal-
       ysis of permit data.

276   many people criticized   La Colectiva, "Issues and Con-
       cerns: Forest Service Policy," policy statement, 6 No-
       vember 1979, El Rito, New Mexico. Also: Ray Stevens,
       Chairman, New Mexico Citizens' Forest Service Ad-
       visory Council, to Jack Crellin and Jim Perry, Super-
       visors of the Carson and Santa Fe National Forests,
       respectively, memorandum, no date (c. 1980).

277   erected roadblocks   Interview with Marty Morrison, Co-
       yote District Ranger, July 1980.

17. Wilderness

279   "In the long run . . ."   As quoted by Flader, *Thinking Like
       a Mountain*, p. 270.

280   herds to extinction   Bailey, *Mammals of the Southwest-
       ern United States*, pp. 39–41. See also, Ligon, *Wild
       Life of New Mexico*, p. 71; Elliot Barker, *Beatty's Cabin*,
       pp. 87–98.

280   tracks . . . last seen   Bailey, pp. 16–17.

280   last of the Pecos grizzlies   Barker, *Beatty's Cabin*, pp. 189–
       90.

281   eight deer were killed   Flader, p. 56, n. 21.

282   state game commission   Ibid., p. 59.

283   rejoin the Forest Service   Ibid., pp. 14–15.

284   enter recreation business   Nash, *Wilderness and the
       American Mind*, pp. 185 ff. Also, Nash, "Historical
       Roots of Wilderness Management," in *Wilderness
       Management*, pp. 33–36.

284   "a continuous stretch . . ."   Aldo Leopold, "Wilderness and
       Its Place in Forest Recreational Policy," p. 719.

285   Pooler approved it   Nash, *Wilderness*, p. 187.

285   redesignated the Pecos Wilderness   U.S.D.A.—Forest Service, "Pecos Wilderness Multiple Use Guide," p. 2.

285   divert an entire river   Interview with Henry Gallegos, former Game Warden and long-time resident of Pecos, New Mexico, 26 May 1976.

287   "no man who reads . . ."   Nash, *Wilderness*, p. 225.

287   popularize the virtues   Ibid., p. 263; Hendee, p. 8.

288   bighorn sheep . . . re-established   Interview with Robert Lange, Wildlife Veterinarian, Department of Game and Fish, 28 May 1976. Also, New Mexico Department of Game and Fish, *New Mexico Wildlife Management*, p. 75. The bighorn sheep transplanted to the Pecos Wilderness in 1964 were taken from a herd in the Sandia Mountains, which in turn had been established in 1940–42 with sheep imported from Banff, Alberta, Canada.

288   Elk . . . restored   Barker, *Beatty's Cabin*, pp. 76–98. The present Pecos elk herd is descended from a band of thirty-seven Yellowstone elk released near Grass Mountain in 1915. Poached and harassed by Terrero miners and other local residents, the herd led a tenuous existence and was not firmly and securely established until the mid-1930s. The present herd consists of Rocky Mountain or Nelson's elk, *Cervus canadensis nelsoni*, which was the species originally native to the Sangres. Another species, Merriam's elk (*C. merriami*), which was native to the mountains of southwestern New Mexico, became extinct about 1900. Much of its former range is now stocked with Rocky Mountain elk (Game and Fish, *Wildlife Management*, p. 31).

        Yet another animal, the white-tailed ptarmigan, was re-introduced to the Pecos Wilderness in 1980 (Interview with Wildlife Staff, Carson Forest Supervisor's Office, June 1981).

289   "green immigrants"   See, Haughton, *Green Immigrants: The Plants that Transformed America.*

289   *manage* the land   Spurr, "Wilderness Management."

290   flames . . . needed again to renew   Hendee, p. 260.

291   "was a long ways . . ."   Interview transcripts for Tucker

and Fitzpatrick, *Men Who Matched the Mountains*, p. 984.

293   three hundred people   Tucker and Fitzpatrick, *Men Who Matched the Mountains*, p. 227.

294   18,598 pleasure campers   Figures on recreational use of the Pecos Wilderness were provided by Joe Quaid, recreation staff, Santa Fe National Forest, Santa Fe. Of National Forest Wilderness Areas comprising 100,000 or more acres, the Pecos Wilderness is one of only five with a visitor-day per acre quotient greater than .75. The other four include the Emigrant Basin, John Muir, and Minarets Wildernesses in California and the Boundary Waters Canoe Area in Minnesota (Hendee, pp. 292–93).

296   Wilderness experience . . . educates   Sax, *Mountains Without Handrails*, pp. 14–15.

18. TRAILS

301   "If these mountains die . . ."   Nichols, *If Mountains Die*, no page number.

304   "secret tunnel"   Interview with Jay Eby, former District Ranger, Española, N.M., July 1976, at Pecos Ranger Station.

304   fortunes at the racetrack   Interview with Elliot Barker, 6 June 1977, Santa Fe.

306   "the Territory of New Mexico . . ."   Albert Pike, *Prose Sketches and Poems*, David J. Weber, ed. (Albuquerque: Calvin Horn, 1967), p. 249.

306   "the emigrant wagon . . ."   Wheeler, *Annual Report*, 1876, p. 129.

308   that they be . . . the very opposite   See John Nichols's eloquent discussion in *If Mountains Die*, pp. 121–41. I differ with Nichols significantly, however, in my assessment of the Forest Service in northern New Mexico.

309   "common life . . ."   Philp, "Albert B. Fall and the Protest from the Pueblos, 1921–1923," p. 248. Philp argues that the controversy provoked by the Bursum Bill may have contributed as much, if not more, than the Teapot

Dome scandal to Fall's decision to resign as Secretary of Interior.

309 667,479 acres . . . returned   Simmons, "History of the Pueblos Since 1821," in Ortiz, *Handbook*, pp. 214-15.

311 Hispanic Land Claims Commission   Cutter, "The Legacy of the Treaty of Guadalupe Hidalgo," p. 314.

313 "provide for public participation . . ."   USDA—Forest Service, *Principal Laws*, p. 347.

314 "Manage the Resources . . ."   USDA—Forest Service, "Proposed Plan," November 1981, p. 10.

316 twenty-seven communes   Gardner, *The Children of Prosperity: Thirteen Modern American Communes*, p. 99.

318 "Sons and Daughters . . ."   Trillin, "U.S. Journal: Santa Fe, N.M."

318 "just another paranoid . . ."   Nichols, *If Mountains Die*, p. 137.

Bibliography

ARCHIVAL MATERIALS

Albuquerque. United States Department of Agriculture—Forest Service. Region 3 Headquarters. Miscellaneous records.

———. University of New Mexico. General Library Special Collections Department. Archive 133. "Frank Bond and Son Business Records."

Santa Fe. State Records Center and Archives. New Mexico Writers' Project. County Files.

———. Records of the Court of Private Land Claims.

———. Santa Fe County District Court Records.

———. Taos County District Court Records.

———. United States Department of the Interior. Bureau of Land Management. Surveyor General Reports. Records of Private Land Claims Adjudicated by the U.S. Surveyor General, 1855–90.

Santa Fe. Museum of New Mexico, History Library. New Mexico Writers' Project Archives.

Santa Fe. Museum of New Mexico. Photographic Archives.

Santa Fe. United States Department of Agriculture—Forest Service. Santa Fe National Forest Supervisor's Office. Miscellaneous Records and Photographs.

Taos. United States Department of Agriculture—Forest Service. Carson National Forest Supervisor's Office. Miscellaneous Records.

BIBLIOGRAPHY

PUBLISHED AND DUPLICATED MATERIALS

Abel, Annie Heloise, ed. *The Official Correspondence of James S. Calhoun While Indian Agent at Santa Fe and Superintendent of Indian Affairs in New Mexico.* Washington: Government Printing Office, 1915.

Adams, Eleanor B., ed. *Bishop Tamaron's Visitation of New Mexico.* Albuquerque: Historical Society of New Mexico, 1954.

Applegate, Frank G. *Native Tales of New Mexico.* Philadelphia: J. B. Lippincott, 1932.

Aragón, Janie Louise. "The People of Santa Fe in the 1790's," *Aztlán* 7 (1976):391–417.

Athearn, Frederic James. "Life and Society in Eighteenth-Century New Mexico, 1692–1776." Ph.D. dissertation, University of Texas, Austin, 1974.

Bailey, Florence Merriam. *Birds of New Mexico.* Santa Fe: New Mexico Department of Game and Fish, 1928.

Bailey, Vernon. *Mammals of the Southwestern United States.* (Orig. title: Mammals of New Mexico.) 1930. Reprint. New York: Dover, 1971.

Bandelier, Adolph F. A. *Final Report of Investigations among the Indians of the Southwestern United States, carried out mainly in the years from 1880 to 1885.* 2 vols. Papers of the Archaeological Institute of America. Cambridge, Mass.: John Wilson & Son, 1892.

———. *The Southwestern Journals of Adolph F. Bandelier, 1880–1882.* Charles H. Lange and Carroll L. Riley, eds. Albuquerque: University of New Mexico Press, 1966. *1883–1884,* 1970.

Bannon, John Francis. *The Spanish Borderlands Frontier, 1513–1821.* 1963. Reprint. Albuquerque: University of New Mexico Press, 1974.

Barker, S. Omar, and Barker, Elsa. "Hermit of the Mountain." *New Mexico Quarterly* 31 (1961–62):349–55.

Bartlett, Richard A. *Great Surveys of the American West.* Norman: University of Oklahoma Press, 1962.

Beck, Warren A., and Haase, Ynez D. *Historical Atlas of New Mexico.* Norman: University of Oklahoma Press, 1969.

Bell, William A. *New Tracks in North America.* 1870. Reprint. Albuquerque: Horn and Wallace, 1965.

Benavides, Alonso de. *The Memorial of Fray Alonso de Benavides, 1630.* Translated by Mrs. Edward E. Ayer. Annotated by Frederick W. Hodge and Charles F. Lummis. 1916. Reprint. Albuquerque: Horn and Wallace, 1965.

Benton, Thomas Hart. *Thirty Years View.* 2 vols. New York: D. Appleton, 1857.

Bidwell, Percy. "The Agricultural Revolution in New England." *American Historical Review* 26 (July 1921):683–702.

Bohrer, Vorsila. "The Prehistoric and Historic Role of the Cool-Season Grasses in the Southwest." *Economic Botany* 29 (1975):199–207.

Bolton, Herbert E. *Coronado, Knight of Pueblos and Plains.* Albuquerque: University of New Mexico Press, 1949.

Bond, Frank. "Memoirs of Forty Years in New Mexico." *New Mexico Historical Review* 21 (1946):340–49.

Bourke, John G. "Bourke on the Southwest." Edited by Lansing B. Bloom. *New Mexico Historical Review* 8–13 (1933–38).

Bowden, Henry Warner. "Spanish Missions, Cultural Conflict, and the Pueblo Revolt of 1680." *Church History* 44 (1975):217–28.

Boyd, E. *Popular Arts of Spanish New Mexico.* Santa Fe: Museum of New Mexico Press, 1974.

Brown, Donald N. "Structural Change at Picuris Pueblo." Ph.D. dissertation, University of Arizona, 1973.

Brown, Lorin, with Briggs, Charles L. and Weigle, Marta. *Hispanic Folklife of New Mexico.* Albuquerque: University of New Mexico Press, 1978.

Bunting, Bainbridge. "An Architectural Guide to Northern New Mexico." *New Mexico Architect* 12 (Sept.–Oct. 1970):13–51.

———. *Early Architecture in New Mexico.* Albuquerque: University of New Mexico Press, 1976.

Burcham, Lee T. "Cattle and Range Forage in California, 1770–1880." *Agricultural History* 35 (1961):140–49.

Cabeza de Baca, Fabiola. *We Fed Them Cactus.* Albuquerque: University of New Mexico Press, 1954.

Calvin, Ross. *Sky Determines.* 1934; rev. ed., Albuquerque: University of New Mexico Press, 1965.

Campa, Arthur L. *Hispanic Culture in the Southwest.* Norman: University of Oklahoma Press, 1979.

———. *Los Comanches: A New Mexico Folk Drama.* University of New Mexico Modern Language Series Bulletin No. 7, 1942.

———. *Treasure of the Sangre de Cristos: Tales and Traditions of the Spanish Southwest.* Norman: University of Oklahoma Press, 1963.

Carlson, Alvar Ward. "New Mexico's Sheep Industry, 1850–1900: Its Role in the History of the Territory." *New Mexico Historical Review* 44 (1969):25–49.

———. "The Rio Arriba: A Geographic Appraisal of the Spanish American Homeland." Ph.D. dissertation, University of Minnesota, 1964.

Carrier, Lyman, and Bort, Katharine S. "The History of Kentucky Bluegrass and White Clover in the United States." *Journal of the American Society of Agronomy* 8 (1916):256–66.

Carroll, H. Bailey, and Haggard, J. Villasana, eds. *Three New Mexico Chronicles: Pino, Barriero, and Escudero.* Albuquerque: Quivira Society, 1942.

Chavez, Fray Angelico. "The Carpenter Pueblo." *New Mexico Magazine* 49 (Sept.–Oct. 1971):26–33.

———. "Early Settlements in the Mora Valley." *El Palacio* 62 (1955): 318–24.

———. *My Penitente Land: Reflections on Spanish New Mexico.* Albuquerque: University of New Mexico Press, 1974.

———. *Origins of New Mexico Families in the Spanish Colonial Period.* Santa Fe: Historical Society of New Mexico, 1954.

———. "The Penitentes of New Mexico." *New Mexico Historical Review* 29 (1954):97–123.

Chittenden, Hiram M. *The American Fur Trade of the Far West.* 3 vols. New York: Francis P. Harper, 1902.

Christiansen, Paige W. *The Story of Mining in New Mexico.* Socorro: New Mexico Bureau of Mines and Mineral Resources, 1974.

———, and Kottlowski, Frank E. *Mosaic of New Mexico's Scenery, Rocks, and History.* Socorro: New Mexico Bureau of Mines and Mineral Resources, 1972.

Cleland, Robert Glass. *This Reckless Breed of Men: The Trappers and Fur Traders of the Southwest.* 1950. Reprint. Albuquerque: University of New Mexico Press, 1976.

Coke, Van Deren. *Taos and Santa Fe: The Artist's Environment.* Albuquerque: University of New Mexico Press, 1963.

Coles, Robert. *The Old Ones of New Mexico.* Albuquerque: University of New Mexico Press, 1973.

———. *Eskimos, Chicanos, Indians.* Boston: Little, Brown, 1977.

Collier, John. *On the Gleaming Way.* 2nd rev. ed. Chicago: Sage Books, 1962.

Collins, Dabney Otis. "Battle for Blue Lake." *American West,* September 1971, pp. 33–37.

Cook, R. E. "Operation Crossties." *New Mexico Magazine,* May 1954.

Cooke, Ronald U., and Reeves, Richard W. *Arroyos and Environmental Change in the American South-West.* Oxford: Clarendon Press, 1976.

Cooper, C. F. "Changes in Vegetation Structure and Growth of Southwestern Pine Forests since White Settlement." *Ecological Monographs* 30 (1960):129–64.

Cooperrider, Charles K., and Hendricks, B. A. "Soil erosion and stream flow on range and forest lands of the Upper Rio Grande watershed region in relation to land resources and human welfare." U.S. Department of Agriculture Technical Bulletin 567, 1937.

Cordova, Lorenzo de (Lorin Brown). *Echoes of the Flute.* Santa Fe: Ancient City Press, 1972.

Cortés, Carlos E., ed. *The New Mexican Hispano.* The Mexican American Series. New York: Arno Press, 1974.

———. *Spanish and Mexican Land Grants.* The Mexican American Series. New York: Arno Press, 1974.

Crosby, Alfred. *The Columbian Exchange: Biological and Cultural Consequences of 1492.* Westport, Conn.: Greenwood Press, 1972.

Curtin, L. S. M. *Healing Herbs of the Upper Rio Grande.* 1947. Reprint. Los Angeles: Southwest Museum, 1974.

Curtis, Edward S. *The North American Indian: Being a Series of Volumes Picturing and Describing the Indians of the United States and Alaska.* Frederick W. Hodge, ed. 20 vols. 1907–30. Reprint. New York: Johnson Reprint, 1970.

Cutter, Donald C. "The Legacy of the Treaty of Guadalupe Hidalgo." *New Mexico Historical Review* 53 (1978):305–15.

Dana, Samuel Trask, and Fairfax, Sally K. *Forest and Range Policy: Its Development in the United States.* 2nd ed., rev. New York: McGraw-Hill, 1980.

de Borhegyi, Stephen F. "The Miraculous Shrines of Our Lord of Esquípulas in Guatemala and Chimayó." *El Palacio* 60 (1953):80–111.

deBuys, William. "Fractions of Justice: A Legal and Social History of the Las Trampas Grant, New Mexico." *New Mexico Historical Review* 56 (1981):71–97.

Denevan, William M. "Livestock Numbers in Nineteenth-Century New Mexico and the Problem of Gullying in the Southwest." *Annals of the Association of American Geographers* 57 (1967): 691–703.

Detwyler, Thomas R., ed. *Man's Impact on Environment*. New York: McGraw-Hill, 1971.

Dickey, Roland F. *New Mexico Village Arts*. Albuquerque: University of New Mexico Press, 1949.

Dickson, D. Bruce, Jr. *Prehistoric Pueblo Settlement Patterns: The Arroyo Hondo, New Mexico, Site Survey*. Arroyo Hondo Archaeological Series, vol. 2. Santa Fe: School of American Research Press, 1979.

Dobie, J. Frank. *Coronado's Children*. 1930. Reprint. Austin: University of Texas Press, 1978.

Domínguez, Francisco Atanasio. *The Missions of New Mexico, 1776*. Translated and annotated by Eleanor B. Adams and Fray Angelico Chavez. Albuquerque: University of New Mexico Press, 1956.

Douglass, William B. "A World-Quarter Shrine of the Tewa Indians." *Records of the Past* 11 (1912):159–73.

Dozier, Edward P. *The Pueblo Indians of North America*. New York: Holt, Rinehart & Winston, 1970.

Dregne, Harold E. "Desertification of Arid Lands." *Economic Geography* 53 (1977):322–31.

Eastman, Clyde; Carruthers, Gary; and Liefer, James A. "Evaluation of Attitudes Toward Land in North-Central New Mexico." New Mexico State University Agricultural Experiment Station Bulletin 577. May 1971.

Eastman, Clyde, and Gray, James R. "Social Analysis Model for the Santa Fe National Forest." Prepared for U.S. Forest Service Contract #43-8379-9-154, Santa Fe, 1979.

Ebright, Malcolm. "The San Joaquín Grant: Who Owned the Common Lands? A Historical-Legal Puzzle." *New Mexico Historical Review* 57 (1982):8–15.

———. *The Tierra Amarilla Grant: A History of Chicanery*. Santa Fe: Center for Land Grant Studies, 1980.

Eckholm, Eric. *Losing Ground: Environmental Stress and World Food Prospects*. New York: W. W. Norton, 1976.

Edwords, Clarence E. *Campfires of a Naturalist*. New York: D. Appleton, 1893.

Ellis, Florence Hawley. "Archaeological History of Nambe Pueblo, 14th Century to the Present." *American Antiquity* 30 (1964):34–42.

———. "Pueblo Boundaries and Their Markers." *Plateau* 38/4 (Spring 1966):97–106.

Ellis, Richard N., ed. *New Mexico Historic Documents.* Albuquerque: University of New Mexico Press, 1975.

————. *New Mexico Past and Present: A Historical Reader.* Albuquerque: University of New Mexico Press, 1971.

Emmett, Chris. *Fort Union and the Winning of the Southwest.* Norman: University of Oklahoma Press, 1965.

Engstrand, Iris Wilson. "Land Grant Problems in the Southwest." *New Mexico Historical Review* 53 (1978):317–36.

Fahl, Ronald J. *North American Forest and Conservation History: A Bibliography.* Santa Barbara, California: Forest History Society, 1977.

Fergusson, Erna. *New Mexico: A Pageant of Three Peoples.* 1966. Reprint. Albuquerque: University of New Mexico Press, 1973.

Fergusson, Harvey. *Rio Grande.* New York: Alfred A. Knopf, 1933.

Ferrán, Manuel Antonio. "Planning for Economic Development of Rural Northern New Mexico, with Emphasis on the Peñasco Valley, Taos County." Ph.D. dissertation, University of Oklahoma, 1969.

Flader, Susan L. *Thinking Like a Mountain: Aldo Leopold and the Evolution of an Ecological Attitude Toward Deer, Wolves, and Forests.* Lincoln: University of Nebraska Press, 1978.

Fowler, Jacob. *The Journal of Jacob Fowler.* Edited by Elliot Coues. 1898. Reprint. Minneapolis: Ross & Haines, 1965.

Friedlander, Eva, and Pinyan, Pamela J. *Indian Use of the Santa Fe National Forest: A Determination from Ethnographic Sources.* Ethnohistorical Report Series No. 1. Albuquerque: Center for Anthropological Studies, 1980.

Fritz, Henry E. "The Cattleman's Frontier in the Trans-Mississippi West: A Critical Bibliography." *Arizona and the West* 14 (1972), nos. 1 & 2.

Frome, Michael. *The Forest Service.* New York: Praeger, 1971.,

Gardner, Hugh. *The Children of Prosperity: Thirteen Modern American Communes.* New York: St. Martin's Press, 1978.

Gardner, Richard. *¡Grito! Reies Tijerina and the New Mexico Land Grant War of 1967.* Indianapolis: Bobbs-Merrill, 1970.

Gibson, Charles. *Spain in America.* New York: Harper & Row, 1966.

Gjevre, John A. *Chili Line: The Narrow Rail Trail to Santa Fe.* Española: Rio Grande Sun Press, 1969.

Goetzmann, William H. *Army Exploration in the American West, 1803–1863.* 1959. Reprint. Lincoln: University of Nebraska Press, 1979.

———. *Exploration and Empire: The Explorer and Scientist in the Winning of the American West.* New York: W. W. Norton, 1966.

———. *When the Eagle Screamed: The Romantic Horizon in American Diplomacy, 1800–1860.* New York: John Wiley & Sons, 1966.

González, Nancie L. *The Spanish Americans of New Mexico: A Heritage of Pride.* Albuquerque: University of New Mexico Press, 1969.

Graber, Linda. *Wilderness as Sacred Space.* Washington: Association of American Geographers, 1976.

Gray, James R. "Economic Benefits from Small Livestock Ranches in North-Central New Mexico." New Mexico State University, Agricultural Experiment Station, Research Report 280. 1974.

Greene, A. C., ed. *The Last Captive: The Lives of Herman Lehmann.* Austin: Encino Press, 1972.

Gregg, Josiah. *The Commerce of the Prairies.* Edited by Milo Milton Quaife. 1925. Reprint. Lincoln: University of Nebraska Press, 1967.

Gross, Frederick A., and Dick-Peddie, William A. "A Map of Primeval Vegetation in New Mexico." *The Southwestern Naturalist* 24 (1979):115–22.

Gross, Frederick A., III. "A Century of Vegetative Change for Northwestern New Mexico." Intern Report, Resources Development Internship Program, Western Interstate Commission for Higher Education, Boulder, Colorado, 1973.

Grubbs, Frank H. "Frank Bond: Gentleman Sheepherder." *New Mexico Historical Review,* vols. 35–37 (1961–62).

Gunnerson, Dolores A. *The Jicarilla Apaches: A Study in Survival.* DeKalb: Northern Illinois University Press, 1974.

Gunnerson, James H. "Apache Archeology in Northeastern New Mexico." *American Antiquity* 34 (1969):23–38.

———. "Archeological Survey in Northeastern New Mexico." *El Palacio* 66/5 (1959):145–54.

Hackett, Charles Wilson, and Shelby, C. C. *Revolt of the Pueblo Indians and Otermin's Attempted Reconquest, 1680–1682.* 2 vols. 1942. Reprint. Albuquerque: University of New Mexico Press, 1970.

Hafen, LeRoy R., ed. *The Mountain Men and the Fur Trade of the Far West.* 10 vols. Glendale, California: Arthur H. Clark, 1965–1972.

Hammond, George P., and Rey, Agapito. *Don Juan de Oñate, Col-*

onizer of New Mexico, 1595–1628. Albuquerque: University of New Mexico Press, 1953.

Hanosh, Eugene J. "A History of Mora." Master's thesis, Highlands University, Las Vegas, New Mexico, 1974.

Hardin, Garret. "The Tragedy of the Commons." *Science* 162 (1968): 1243–48.

Harper, Allan G.; Oberg, Kalvero; and Cordova, Andrew. *Man and Resources in the Middle Rio Grande Valley.* Albuquerque: University of New Mexico Press, 1943.

Harrington, John Peabody. "The Ethnogeography of the Tewa Indians." *29th Annual Report of the Bureau of American Ethnology for the Years 1907–1908.* Washington: Government Printing Office, 1916, pp. 29–636.

Haskett, Bert. "Early History of the Cattle Industry in Arizona." *Arizona Historical Review* 6 (1935):3–42.

Hassell, Jean M. "The People of Northern New Mexico and the National Forests." Unpublished report, U.S. Forest Service, Regional Office, Albuquerque, 13 May 1968.

Haughton, Claire Shaver. *Green Immigrants: The Plants that Transformed America.* New York: Harcourt Brace Jovanovich, 1978.

Hawley, J. W., ed. *Guidebook to Rio Grande Rift in New Mexico and Colorado.* Socorro: New Mexico Bureau of Mines and Mineral Resources, 1978.

Hayes, Alden C. *The Four Churches of Pecos.* Albuquerque: University of New Mexico Press, 1974.

Hays, Samuel P. *Conservation and the Gospel of Efficiency.* 1959. Reprint. New York: Atheneum, 1975.

Hende, John C.; Stankey, George H.; and Lucas, Robert C. *Wilderness Management.* USDA-Forest Service, Miscellaneous Publication No. 1365. Washington: Government Printing Office, 1978.

Henderson, Alice Corbin. *Brothers of Light.* New York: Harcourt Brace, 1937.

Henderson, Charles W. *Gold, Silver, Copper, Lead and Zinc in New Mexico and Texas in 1928.* Washington: U.S. Department of Commerce, 1930.

Henderson, Harold G. *An Introduction to Haiku.* Garden City, N.Y.: Doubleday Anchor, 1958.

Hewett, Edgar L. *Ancient Life in the American Southwest.* Indianapolis: Bobbs-Merrill, 1930.

———. "Studies on the Extinct Pueblo of Pecos." *American Anthropologist* 6 (1904):426–39.

————, and Dutton, Bertha P. *The Pueblo Indian World: Studies on the Natural History of the Rio Grande Valley in Relation to Pueblo Indian Culture.* Albuquerque: University of New Mexico Press, 1945.

Hitchcock, A. S. *Manual of the Grasses of the United States.* 2d ed., rev. by Agnes Chase, 1950. Reprint (2 vols.). New York: Dover, 1971.

Hodge, Carle, ed. *Aridity and Man: The Challenge of the Arid Lands in the United States.* Washington: American Association for the Advancement of Science, 1963.

Holmes, Kenneth L. *Ewing Young: Master Trapper.* Portland, Oregon: Peter Binford Foundation, 1967.

Horgan, Paul. *The Centuries of Santa Fe.* New York: E. P. Dutton, 1956.

————. *Great River: The Rio Grande in North American History.* 2 vols. New York: Holt, Rinehart & Winston, 1954.

————. *Lamy of Santa Fe, His Life and Times.* New York: Farrar, Straus, and Giroux, 1975.

Hulbert, Archer Butler, ed. *Southwest on the Turquoise Trail: The First Diaries on the Road to Santa Fe.* Denver: Denver Public Library, 1933.

Irwin-Williams, Cynthia. "The Oshara Tradition: Origins of Anasazi Culture." *Eastern New Mexico University Contributions in Anthropology* 5/1 (1973).

Jenkins, Myra Ellen. "History of Las Trampas." Mimeographed, c. 1975.

————. "Spanish Land Grants in the Tewa Area." *New Mexico Historical Review* 47 (1972):113–34.

Jenkinson, Michael. *Ghost Towns of New Mexico.* Albuquerque: University of New Mexico Press, 1967.

John, Elizabeth. *Storms Brewed in Other Men's Worlds.* College Station, Texas: Texas A & M Press, 1975.

Johnson, John W. *Reminiscences of a Forest Ranger, 1914–1944.* Dayton, Ohio: Joan Canby Robinette, 1976.

Jones, Fayette. *Old Mining Camps of New Mexico, 1854–1904.* 1904. Reprint. Santa Fe: Stagecoach Press, 1964.

Keleher, William A. "Law of the New Mexico Land Grant." *New Mexico Historical Review* 4 (1929):350–71.

————. *Turmoil in New Mexico, 1846–1868.* Santa Fe: Rydal Press, 1952.

Kelley, N. Edmund. *The Contemporary Ecology of Arroyo Hondo, New Mexico.* Arroyo Hondo Archaeological Series, vol. 1. Santa Fe: School of American Research Press, 1980.

Kenner, Charles L. *A History of New Mexico—Plains Indians Relations.* Norman: University of Oklahoma Press, 1969.

Kessell, John L. *Kiva, Cross, and Crown.* Washington: National Park Service, 1979.

————. *The Missions of New Mexico Since 1776.* Albuquerque: University of New Mexico Press, 1980.

Knowlton, Clark S. "Causes of Land Loss Among the Spanish Americans in Northern New Mexico." *Rocky Mountain Social Science Journal* 1 (1963):201–11.

————. "Changing Spanish Villages of Northern New Mexico." *Sociology and Social Research* 53 (1969):455–74.

————. "Violence in New Mexico." *California Law Review* 58 (1970): 1054–85.

Kutsche, Paul, ed. *The Survival of Spanish American Villages,* Colorado College Studies, No. 15. Spring, 1979.

Kvasnicke, Robert M., and Viola, Herman J., eds. *The Commissioners of Indian Affairs, 1824–1977.* Lincoln: University of Nebraska Press, 1979.

Lamar, Howard Roberts. *The Far Southwest, 1846–1912: A Territorial History.* New Haven: Yale University Press, 1966.

Lang, E. M. *Deer of New Mexico.* Santa Fe: New Mexico Department of Game and Fish, 1957.

————. *Elk of New Mexico.* Santa Fe: New Mexico Department of Game and Fish, 1958.

Lange, Charles H. *Cochiti: A New Mexico Pueblo, Past and Present.* 2nd ed. rev. Carbondale and Edwardsville: Southern Illinois University Press, 1968.

Lavender, David. *Bent's Fort.* Garden City, N.Y.: Doubleday, 1954.

————. *The Southwest.* New York: Harper & Row, 1980.

Leaf, Charles Frank. "Watershed Management in the Central and Southern Rocky Mountains: a summary of the status of our knowledge by vegetation types." U.S. Forest Service Research Paper RM-142. March 1975.

Le Noir, Phil. "The Hermit of Las Vegas." *Publications of the Texas Folklore Society* 10 (1932):124–26.

Leonard, Olen Earl. *The Role of the Land Grant in the Social Organization and Social Processes of a Spanish-American Village in New Mexico.* 1943. Reprint. Albuquerque: Calvin Horn, 1970.

Leopold, Aldo. *A Sand County Almanac, and Sketches Here and There.* New York: Oxford University Press, 1949.

———. "Wilderness and Its Place in Forest Recreational Policy." *Journal of Forestry* 19 (1921):718–21.

Leopold, Luna B. "Vegetation of Southwestern Watersheds in the Nineteenth Century." *Geographical Review* 41 (1951):295–316.

Lewis, Henry T. "Indian Fires of Spring." *Natural History* 89 (January 1980):76–83.

Ligon, J. Stokely. *Wild Life of New Mexico.* Santa Fe: Department of Game and Fish, 1927.

López, Larry. "The Founding of San Francisco on the Rio Puerco: A Document." *New Mexico Historical Review* 55 (January 1980):71–78.

———. *Taos Valley: A Historical Survey.* Boulder, Colorado: Western Interstate Commission for Higher Education, 1975.

Lummis, Charles F. *The Land of Poco Tiempo.* 1893. Reprint. Albuquerque: University of New Mexico Press, 1966.

McCarty, Frankie. "Land Grant Problems in New Mexico." Albuquerque *Journal*, Sept. 28–Oct. 10, 1969. Reprint (pamphlet). Albuquerque: Albuquerque *Journal*, 1969.

Malde, Harold E., and Scott, Arthur G. "Observations of Contemporary Arroyo Cutting Near Santa Fe, New Mexico, U.S.A." *Earth Surface Processes* 2 (1977):39–54.

Marsh, George Perkins. *Man and Nature, Or, Physical Geography as Modified by Human Action.* Edited by David Lowenthal. 1864. Reprint. Cambridge: Harvard University Press, Belknap Press, 1965.

Mayhew, Ann. "A Reappraisal of the Causes of Farm Protest in the U.S., 1870–1900." *Journal of Economic History* 32 (June 1972):464–74.

Mead, Margaret, ed. *Cultural Patterns and Technical Change.* New York: UNESCO, 1953.

Meinig, D. W. *Southwest: Three Peoples in Geographical Change, 1600–1970.* New York: Oxford University Press, 1971.

Miller, John P.; Montgomery, Arthur; and Sutherland, Patrick K. *Geology of Part of the Southern Sangre de Cristo Mountains.* Socorro: New Mexico Bureau of Mines and Mineral Resources, 1963.

Mollin, Ferdinand E. *If and When It Rains: The Stockman's View of the Range Question.* Denver: American National Livestock Association, 1938.

Momaday, N. Scott. *House Made of Dawn.* 1968. New York: Signet, 1969.

Moorhead, Max L. *New Mexico's Royal Road: Trade and Travel on the Chihuahua Trail.* Norman: University of Oklahoma Press, 1958.

Myers, John M., compiler and editor. *The Westerner: A Roundup of Pioneer Reminiscences.* Englewood Cliffs, N.J.: Prentice-Hall, 1969.

Nabokov, Peter. *Tijerina and the Courthouse Raid.* 2nd ed., rev. Albuquerque: University of New Mexico Press, 1970.

Nash, Roderick. *Wilderness and the American Mind.* 2nd ed., rev. New Haven: Yale University Press, 1973.

New Mexico Bureau of Immigration. *The Mines of New Mexico: Inexhaustible Deposits of GOLD AND SILVER, Copper, Lead, Iron, and Coal.* Santa Fe: New Mexico Bureau of Immigration, 1896.

New Mexico Department of Game and Fish. *New Mexico Game Management.* Santa Fe: New Mexico Department of Game and Fish, 1967.

New Mexico Geological Society. *Guidebook of Southeastern Sangre de Cristo Mountains, New Mexico.* Seventh Field Conference, 19–21 October 1956. Socorro: New Mexico Geological Society, 1956.

Nichols, John. *If Mountains Die.* New York: Alfred A. Knopf, 1979.

Northrop, Stuart A. *Minerals of New Mexico.* Albuquerque: University of New Mexico Press, 1942.

Nostrand, Richard L. "The Hispanic-American Borderland: A Regional, Historical Geography." Ph.D. dissertation, University of California, Los Angeles, 1968.

Opler, Morris Edward. "Jicarilla Apache Territory, Economy, and Society in 1850." *Southwestern Journal of Anthropology* 27 (1971):309–29.

———. *Myths and Tales of the Jicarilla Apache Indians.* New York: American Folk-lore Society, 1938.

Ortiz, Alfonso, ed. *New Perspectives on the Pueblos.* Albuquerque: University of New Mexico Press, 1972.

———, ed. *Southwest. Handbook of North American Indians,* vol. 9. William C. Sturtevant, general editor. Washington: Smithsonian Institution, 1979.

————. *The Tewa World: Space, Time, Being, and Becoming in a Pueblo Society.* Chicago: University of Chicago Press, 1969.

Ozarks—Four Corners Regional Development Commission. Hearings before a Special Subcommittee of the Committee on Public Works. U.S. Senate, 90th Congress, 1st Session. Espanola, Taos, Las Vegas, New Mexico, August 1967.

Pancoast, Charles Edward. *A Quaker Forty-Niner.* Ed. by A. P. Hannum. Philadelphia: University of Pennsylvania Press, 1930.

Paredes, Raymond A. "The Mexican Image in American Travel Literature, 1831–69." *New Mexico Historical Review* 52 (1977):5–29.

Parish, William J. *The Charles Ilfeld Company: A Study of the Rise and Decline of Mercantile Capitalism in New Mexico.* Cambridge, Mass.: Harvard University Press, 1961.

————. "Sheep Husbandry in New Mexico, 1902–1903." *New Mexico Historical Review* 37–38 (1961–62).

Parkman, Francis. *The Oregon Trail.* 1849. Edited by E. N. Feltskog. Madison: University of Wisconsin Press, 1969.

Parr, John M. "The Development and Movement of Tree Islands Near the Upper Limit of Tree Growth in the Southern Rocky Mountains." *Ecology* 58 (1977):1159–64.

Parsons, Elsie Clews. *Pueblo Indian Religion.* 2 vols. Chicago: University of Chicago Press, 1939.

Pearce, T. M., ed. *New Mexico Place Names.* Albuquerque: University of New Mexico Press, 1965.

Perrigo, Lynn I. *The American Southwest: Its Peoples and Cultures.* New York: Holt, Rinehart & Winston, 1971.

————. *Gateway to Glorieta: A History of Las Vegas, New Mexico.* Boulder, Colo.: Pruett Publishing Co., 1982.

————. "The Original Las Vegas 1835–1935." Typescript. Carnegie Public Library, Las Vegas, New Mexico, c. 1975.

Philp, Kenneth R. "Albert B. Fall and the Protest from the Pueblos, 1921–1923." *Arizona and the West* 12 (1970):237–54.

————. *John Collier's Crusade for Indian Reform, 1920–1954.* Tucson: University of Arizona Press, 1977.

Pike, Albert. *Prose Sketches and Poems.* Edited by David J. Weber. Albuquerque: Calvin Horn, 1967.

Pinchot, Gifford. *Breaking New Ground.* New York: Harcourt Brace, 1947.

Powell, John Wesley. *Report on the Lands of the Arid Region of the*

United States. 2nd ed., rev. Washington: Government Printing Office, 1879.

Pulling, Hazel A. "Range Forage and California's Range Cattle Industry." *Historian* 7 (1945):113–29.

Ray, L. L. "Glacial Chronology of the Southern Rocky Mountains." *Geological Society of America Bulletin* 51 (1940):1851–1918.

Reeve, Frank D. "The Sheep Industry in Arizona, 1903–1906." *New Mexico Historical Review* 38–39 (1962–63).

Reinhart, Theodore R. "Late Archaic Cultures of the Middle Rio Grande Valley, New Mexico." Ph.D. dissertation, University of New Mexico, 1968.

Reno, Philip. *And Farther On Was Gold in Twining, New Mexico.* Denver: Sage, 1962.

Ríos-Bustamante, Antonio José. "New Mexico in the Eighteenth Century: Life, Labor, and Trade in *la Villa de San Felipe de Albuquerque.*" *Aztlán* 7 (1976):357–89.

Roberts, Paul H. *Hoofprints on Forest Ranges: The Early Years of National Forest Range Administration.* San Antonio: Naylor, 1963.

Robinson, Glen O. *The Forest Service: A Study in Public Land Management.* Baltimore: Resources for the Future, 1975.

Rosenbaum, Robert J. "*Mexicano* vs. *Americano:* A Study of Hispanic-American resistance to Anglo-American control in New Mexico Territory, 1870–1900." Ph.D. dissertation, University of Texas at Austin, 1972.

Ross, Ivy B. *The Confirmation of Spanish and Mexican Land Grants in California.* San Francisco: R and E Research Associates, 1974.

Ruxton, George Frederick Augustus. *Ruxton of the Rockies.* Edited by LeRoy R. Hafen. Norman: University of Oklahoma Press, 1950.

Sax, Joseph L. *Mountains Without Handrails: Reflections on the National Parks.* Ann Arbor: University of Michigan Press, 1980.

Schery, Robert W. "The Migration of a Plant: Kentucky Bluegrass Followed Settlers of New World." *Natural History* 74 (1965):41–49.

Schroeder, Albert H. "Shifting for Survival in the Spanish Southwest." *New Mexico Historical Review* 43 (1968):291–310.

Scully, Vincent. *Pueblo: Mountain, Village, Dance.* New York: Viking, 1975.

Servín, Manuel P., ed. *An Awakened Minority: The Mexican Americans.* Los Angeles: Glencoe, 1974.

Shanin, Teodor. *Peasants and Peasant Societies*. Harmondsworth, Middlesex: Penguin Books, 1971.

Shepherd, Jack. *The Forest Killers: The Destruction of the American Wilderness*. New York: Weybright and Talley, 1975.

Sheridan, David. *Desertification of the United States*. Washington: Council on Environmental Quality, 1981.

Sherman, James E., and Sherman, Barbara H. *Ghost Towns and Mining Camps of New Mexico*. Norman: University of Oklahoma Press, 1975.

Simmons, Marc. *New Mexico: A History*. New York: W. W. Norton, 1977.

——. "Settlement Plans and Village Patterns in Colonial New Mexico." *Journal of the West* 8 (1969):7–21.

——. *Two Southwesterners: Charles Lummis and Amado Chaves*. Cerillos, N.M.: San Marcos Press, 1968.

Smith, E. R. "History of Grazing Industry and Range Conservation in the Rio Grande Basin." *Journal of Range Management* 6 (1953): 405–9.

Spurr, Stephen H. "Wilderness Management." Horace M. Albright Conservation Lectureship, No. 6. University of California School of Forestry. Berkeley, 3 May 1960.

Starr, Frederick. "Shrines Near Cochiti." *American Antiquarian and Oriental Journal* 22 (1900):219–31.

Steen, Harold K. *The U.S. Forest Service: A History*. Seattle: University of Washington Press, 1976.

Steiner, Stan. *La Raza: The Mexican Americans*. New York: Harper & Row, 1970.

Stewart, Omar C. "Forest and Grass Burning in the Mountain West." *Southwestern Lore* 20 (1954):42–46, 59–64; 21 (1955): 5–8.

Stewart, Peter. "Watershed Characteristics Part B. Decreases in Flow in the Rio Grande." Unpublished report. Carson National Forest, Supervisor's Office. Taos, N.M., March 1981.

Stewart, Tom. "Blazing Trail on the Pecos." *New Mexico Magazine*, March 1942.

Sutherland, Patrick K., and Montgomery, Arthur. *Trail Guide to Geology of the Upper Pecos*. 3rd ed., rev. Socorro: New Mexico Bureau of Mines and Mineral Resources, 1975.

Swadesh, Frances Leon. *Los Primeros Pobladores: Hispanic Americans of the Ute Frontier*. Notre Dame: University of Notre Dame Press, 1974.

————. *20,000 Years of History: A New Mexico Bibliography.* Santa Fe: Sunstone, 1973.

————; Vigil, Julian W.; and Ochoa, Marina B. "The Lands of New Mexico." Bilingual Teacher Training Unit, Division of Research, Planning and Innovation, State Department of Education. March 1976.

Taylor, Morris F. "Campaigns Against the Jicarilla Apaches, 1854." *New Mexico Historical Review* 44 (1969):269–91.

Thomas, Alfred Barnaby. *After Coronado: Spanish Exploration Northeast of New Mexico, 1696–1727.* Norman: University of Oklahoma Press, 1935.

————. *Forgotten Frontiers: A Study of the Spanish Indian Policy of Don Juan Bautista de Anza, Governor of New Mexico, 1777–1787.* Norman: University of Oklahoma Press, 1932.

————. *The Plains Indians and New Mexico, 1751–1778.* Albuquerque: University of New Mexico Press, 1940.

Tobias, Henry J., and Woodhouse, Charles E., eds. *Minorities and Politics.* Albuquerque: University of New Mexico Press, 1969.

Trejo, Arnulfo D., ed. *The Chicanos: As We See Ourselves.* Tucson: University of Arizona Press, 1979.

Trillin, Calvin. "U.S. Journal: Santa Fe, N.M." *The New Yorker,* 29 March 1982, pp. 124–27.

Tuan, Yi-Fu. "New Mexican Gullies: A Critical Review and Some Recent Observations." *Annals of the Association of American Geographers* 56 (1966):573–97.

————; Everard, Cyril E.; and Widdison, Jerold G. *The Climate of New Mexico.* Santa Fe: New Mexico State Planning Office, 1969.

Tucker, Edwin A., and Fitzpatrick, George. *Men Who Matched the Mountains: The Forest Service in the Southwest.* Washington: Government Printing Office, 1972.

————, compilers. Interview transcripts for *Men Who Matched the Mountains.* Unpublished ms, c. 1970. U.S.D.A.—Forest Service, Regional Office, Albuquerque.

Twitchell, Ralph Emerson. *The Leading Facts of New Mexican History.* 5 vols. Cedar Rapids: Torch Press, 1911–1917.

————. *Spanish Archives of New Mexico.* 2 vols. Cedar Rapids: Torch Press, 1914.

Tyler, Daniel. "Anglo-American Penetration of the Southwest." *Southwestern Historical Quarterly* 75 (1972):325–38.

Ungnade, Herbert E. "Archeological Finds in the Sangre de Cristo Mountains of New Mexico." *El Palacio* 80 (1963):15–20.

——. *Guide to the New Mexico Mountains.* Albuquerque: University of New Mexico Press, 1965.

United Nations. *United Nations Conference on Desertification: Roundup, Plan of Action, and Resolutions.* New York: United Nations, 1978.

United States, Bureau of Census. *Sixteenth, Seventeenth, Eighteenth, and Nineteenth Census of the United States.* Washington: Bureau of the Census, 1942–1973.

United States, Congress. "Col. Price's Report." Executive Document No. 8, Item 13, 30th Cong., 1st Sess., 1847.

——. "Insurrection Against the Military Government in New Mexico and California, 1847 and 1848." Senate Document No. 442, 56th Cong., 1st Sess., 1900.

United States, Department of Agriculture—Forest Service. Carson and Santa Fe National Forests. "Pecos Wilderness Management Plan." Taos and Santa Fe, 1972.

——. "Pecos Wilderness Multiple Use Guide." Taos and Santa Fe, 1972.

——. *The Principal Laws Relating to Forest Service Activities.* Agriculture Handbook No. 453. Washington: Government Printing Office, 1978.

——. "Proposed Santa Fe National Forest Plan." November 1981.

——. "The Western Range." U.S. Senate Doc. 199, 74th Congress, 2nd Session, 1936.

——. Department of Agriculture—Soil Conservation Service. "Notes on community-owned land grants in New Mexico." Section of Human Surveys—SCS, Region 8. August 1937.

Utley, Robert M. *Fort Union.* Washington: National Park Service, 1962.

van Dresser, Peter. *A Landscape for Humans.* Santa Fe: Lightning Tree, 1972.

Van Ness, John R. "Spanish-American vs. Anglo-American Land Tenure and the Study of Economic Change in New Mexico." *Social Science Journal* 13/3 (1976):45–52.

Villagrá, Gaspar Pérez de. *History of New Mexico.* Translated by Gilberto Espinosa. Los Angeles: Quivira Society, 1933.

Voigt, William. *Public Grazing Lands.* New Brunswick, N.J.: Rutgers University Press, 1976.

Wagoner, J. J. "Overstocking of the Ranges in Southern Arizona During the 1870's and 1880's." *Arizoniana* 2 (1961):23–27.

Wallace, Anthony F. C. *Culture and Personality.* New York: Random House, 1961.

———. *The Death and Rebirth of the Seneca.* New York: Alfred A. Knopf, 1970.

Wallace, William Swilling, ed. *A Journey Through New Mexico's First Judicial District in 1864.* Los Angeles: Westernlore Press, 1956.

Warner, Ted J., ed. *The Domínguez-Escalante Journal.* Translated by Fray Angelico Chavez. Provo, Utah: Brigham Young University Press, 1976.

Weber, David J., ed. *Foreigners in Their Native Land: Historical Roots of the Mexican Americans.* Albuquerque: University of New Mexico Press, 1973.

———. "Mexico's Far Northern Frontier, 1821–1848: A Critical Bibliography." *Arizona and the West* 19 (1977):225–66.

———, ed. *New Spain's Far Northern Frontier.* Albuquerque: University of New Mexico Press, 1979.

———. *The Taos Trappers: The Fur Trade in the Southwest, 1540–1846.* Norman: University of Oklahoma Press, 1971.

Weber, William A. *Rocky Mountain Flora.* Boulder: Colorado Associated University Press, 1976.

Weigle, Marta. *Brothers of Light, Brothers of Blood: The Penitentes of the Southwest.* Albuquerque: University of New Mexico Press, 1976.

———, ed. *Hispanic Villages of Northern New Mexico. A Reprint of Volume II of the 1935 Tewa Basin Study, with Supplementary Materials.* Santa Fe: Lightning Tree, 1975.

Wendorf, Fred. "The Archaeology of Northeastern New Mexico." *El Palacio* 67 (April 1960):55–65.

———, and Miller, John P. "Artifacts from High Mountain Sites in the Sangre de Cristo Range, New Mexico." *El Palacio* 66 (April 1959):37–52.

Wentworth, Edward Norris. *America's Sheep Trails.* Ames, Iowa: Iowa State College Press, 1948.

Westphall, Victor. *The Public Domain in New Mexico, 1854–1891.* Albuquerque: University of New Mexico Press, 1965.

———. *Thomas Benton Catron and His Era.* Tucson: University of Arizona Press, 1973.

Wetherington, Ronald K. *Excavations at Pot Creek Pueblo.* Fort Burgwin Research Center Publication 6. Taos, N.M., 1968.

Wheat, Carl I. *Mapping the Transmississippi West, 1540–1861.* 5 vols. in 6. San Francisco: Institute of Historical Cartography, 1957–1963.

Wheeler, George Montague. *Annual Report upon the Geographical Surveys West of the One Hundredth Meridian.* Washington: Government Printing Office, 1873–84.

———. *Report upon the United States Geographical Surveys West of the One Hundredth Meridian.* Washington: Government Printing Office, 1889.

White, Gilbert F. *The Future of Arid Lands.* Washington: American Association for the Advancement of Science, 1956.

White, Koch, Kelley and McCarthy, Attorneys at Law and the New Mexico Planning Office. *Land Title Study.* Santa Fe: State Planning Office, 1971.

Williams, Jerry L., and McAllister, Paul E., eds. *New Mexico in Maps.* 1979. Reprint. Albuquerque: University of New Mexico Press, 1981.

Wilson, Iris H. *William Wolfskill.* Glendale, Calif.: Arthur H. Clark, 1965.

Wilson, Lee, and Associates. "Future Water Issues, Taos County, New Mexico." Unpublished report, December 1980. Santa Fe.

Wooten, Elmer Otis. "The Range Problem in New Mexico," New Mexico Agricultural Experiment Station Bulletin no. 66, April 1908.

Wormington, H. M. *Prehistoric Indians of the Southwest.* Denver: Denver Museum of Natural History, 1947.

Zeleny, Carolyn. *Relations Between the Spanish-Americans and the Anglo-Americans in New Mexico.* New York: Arno Press, 1974.

Index

abandonments, 18, 44, 58, 59

Abiquiu, N.M., 65, 69, 115, 259

Acoma Pueblo, 17; and Oñate's colonists, 48

Act of March 3, 1891, 237

adobe, 195–96

agriculture, 31–32, 215, 216, 217–18, 219, 230. *See also* Hispanic villagers; subsistence agriculture

Agostini, Giovanni, 122, 135–45, 147

Albuquerque, N.M., 15, 21, 70, 107, 179, 228, 262, 263, 276, 294

Alianza Federal de las Mercedes, 258, 260, 261–62, 275; Courthouse Raid, 261–62, 275, 277; today, 261

All Pueblo Council, 309

alpine fir, 35

alpine zone, 36–38

American traders, 87–100, 106, 114; competed with Hispanos for Mexican trade, 99–100

Americans: alienated from Mexicans during 1900s, 99; racist, 101; relationship with Hispanos during 1800s, 100–101, 104, 127

American Metal Company, 155–57

Anasazi, 29, 38–39; ruins, 165

Anglos, 147, 273–74, 304, 305; and U.S. Forest Service, 246; buying land grants, 171–92, 307; expansion into New Mexico, 11, 111–12, 124; Los Alamos, 316; prospectors, 147–60; relationship with Hispanos, 127–28, 131–32, 163, 168, 190, 193, 261, 306–7,